THE SOVIET AIR FORCE

Also by Asher Lee:

THE GERMAN AIR FORCE. 1946 (Out of print)

AIR POWER. 2nd Edition, 1957

BLITZ ON BRITAIN. 1960

THE SOVIET AIR FORCE

by

ASHER LEE, O.B.E.

GERALD DUCKWORTH & CO. LTD.
3 HENRIETTA STREET, LONDON, W.C.2

First published in 1950
Second Edition 1952
Revised and enlarged Edition 1961

Printed in Great Britain by
The Camelot Press Ltd., London and Southampton

CONTENTS

LIST OF ILLUSTRATIONS

For

MERVYN HORDER

INTRODUCTION

SINCE the outbreak of the Korean War in the summer of 1950, much reliable information has been published on the subject of Soviet air power. There has been unexpected public display of some of the latest Soviet jet fighters, bombers and passenger transport planes and also of Soviet helicopters. The testimony of large numbers of German aircraft engineers and other scientists back after working in the USSR for a decade or more, coupled with the evidence of pilots, radar operators and aircraft mechanics expatriated from Poland, Hungary and Czechoslovakia, has combined to give the student of Soviet air power a series of tangible reference points enabling him to trace the remarkable series of post-war developments with some accuracy.

It is fairly clear that the Soviet Union has, or is about to possess, the complete range of modern air and rocket weapons enabling the Kremlin to launch and fight a strategic thermo-nuclear war across the oceans and land masses or to deal with the contingencies of local limited wars of either the nuclear or non-nuclear type. It is hard to think of any important branch of modern air and missile power in which the USSR has not achieved or could not quickly achieve at least parity with the Western powers. It is a reasonable assessment that in the development of long-range rockets the Soviet Union has an advantage over the United States and that in the next two or three years this advantage in strategic air and missile attack could become preponderant.

Why then does one subscribe to the view that the Soviet Union will not of its own volition make war on the West and that its policy is "all mischief short of war"? This was the view expressed in an assessment of Soviet policy published by the British Royal Institute of International Affairs, Chatham House, London, on the eve of the Korean War in 1950 in a publication called "Defence in the Cold War". Nor was this an exclusively British viewpoint. Two or three years previously in *The Price of Power*, based on the views of the U.S. Council of Foreign Relations (Harper, 1947), Hanson

Baldwin a leading U.S. strategist wrote: "The men in the Kremlin do not want war, if they can achieve their aims without it. . . . The theory of the inevitability of the eventual fall of capitalism has the fortunate connotation that there is no hurry about it."

When the Nato organisation was formed in 1949 there was a widely held view that the middle 1950s would be a danger point for war between East and West. Now the view has been circulated and is gaining currency that there is a danger that, in the middle 1960s, Soviet technical advantages in long-range nuclear and thermonuclear rockets may be so great that she will be in a position to launch a surprise or pre-emptive attack in which she can neutralise the major part of the Western bomber and rocket deterrent forces. Certainly there were many essays in brinkmanship by both East and West in the 1950s, but in each case, while the Soviet leadership through its press and radio breathed fire and brimstone, uttering threats of rocket devastation or branding the Western powers as warmongers and peace-breakers, the important fact is that the USSR kept its powder dry or virtually dry. It interfered directly in Korea much less than it had in the Spanish Civil War; having supplied China with weapons and political encouragement to defeat the forces of Chiang Kai-shek the Kremlin has been cautious almost to the point of discouraging China's desire to attack Formosa; the same Soviet prudence has been noted in the case of the Chinese conquest of Tibet and Pekin's incursions on the frontiers of India. And in the Suez campaign in the autumn of 1956, Soviet radio announcements of Communist volunteers pouring into Egypt in support of Nasser from China and elsewhere and the reported threat to destroy London and Paris with long-range rockets were in fact a verbal fog screen to hide Soviet discretion in ordering her pilots and air technicians in Egypt to fly away from the dangers of the battle zone to Syria and the Sudan. Nor can the Soviet Union be said to have exploited other Middle East situations in a manner suggesting that military action is in the forefront of Soviet policy. Little or no attempt was made to aggravate the incidents on the Turkish-Syrian border or to react, except by radio and press propaganda, when a crisis in Jordan, Lebanon and Irak provoked the intervention of British and American troops in the first two territories. Indeed the entire Soviet policy towards Persia, the United Arab Republic and Irak in the last five years demonstrates that Soviet foreign policy intends to win its victories by propaganda and cultural, political and economic infiltration rather than by military action. Red Army operations in Hungary in the autumn of 1956 were no doubt no more aggressive in the Soviet viewpoint than French action in Algeria or British action in Cyprus or East Africa. Not that the comparison can be made, strictly speaking. But in all cases the major powers concerned no doubt feel that the affair was largely domestic.

But if the 1950s showed no design of major military aggression by the Kremlin, have we any right to assume that the pattern will continue in the 1960s? To what extent can one believe that the disarmament proposals made by the Kremlin in the last five or six years, which have admitted the principle of inspection of Soviet territory, are genuine, and that, in the nuclear and thermo-nuclear age of rockets the Soviet Union genuinely wishes to disarm under international control? In reiterating the view that the Soviet disarmament proposals are genuine, one does not lose sight of the fact, that they will at every step and turn of the inevitably protracted negotiations seek to gain some political advance (e.g. the admission of Communist China to the United Nations) or establish inspection conditions favourable to the Kremlin.

There is reason to believe that in the great revolution in military thinking, military weapons and tactics which has taken place in the USSR in the last decade there has been a growing realisation that the bulk of present weapons are untried in war and that there is nowadays no way of either ensuring or reasonably forecasting the success or pattern of any military plan. Never before in the annals of military history has it been more difficult to formulate a defence policy. It is indeed ironic that in 1960 Soviet military policy is orientated more towards relying on a Soviet deterrent with both strategic and tactical missiles, allied to a proposal to cut its ground forces, at a time when many people in the West, which has for so long relied on the nuclear deterrent, are advocating a strengthening of its conventional ground forces. The proposed cuts in the Soviet armed forces' strength are of course no indication in themselves that Soviet disarmament proposals are genuine. They in no sense constitute a reduction in Soviet military strength. Only a thorough inspection procedure could confirm such a reduction.

The main military reason for the Soviet disarmament proposals is that they find themselves in the same kind of military dilemma that has plagued Western military policy makers for the last ten years. At one end of the scale they can never be certain when a minor military engagement between East and West German forces, or on the Indian border or in Irak, will not explode rapidly into a large-scale nuclear world war of which the outcome is almost bound to be very damaging to the USSR and the final result uncertain. At the other end of the scale there is the dangerous uncertainty of an all-out surprise or pre-emptive mass thermo-nuclear attack. For this to be a success, large numbers of Soviet long-range rockets must be carefully sited and fuelled beforehand and a carefully synchronised simultaneous launching time achieved without prior knowledge of either the timing or the plan reaching a prospective enemy. At a time when more and more tourists, diplomats and technicians and journalists are roaming about the Soviet Union and when large numbers of Arabs, Chinese, Africans, Latin Americans and others are in the country, the risk of a security leak is much greater than five years ago. And whatever successes the USSR may have in making its codes indecipherable, any large plan of major rocket assault must involve thousands of communications by ordinary links, some of which are vulnerable to local carelessness and therefore to interception. One disloyal member of the Soviet codes and ciphers staff, one senior officer whose conscience is pricked, one careless encoder, one lost document could be enough to compromise the plan. And the consequences of such a compromise could be disastrous for the USSR either politically or militarily or both. And added to the uncertainty of security is the uncertainty of the effect of their strategic deterrent forces. However accurate their long-range rockets may be when fuelled, serviced and launched under ideal conditions of peace and by élite personnel, no Soviet planner can estimate their accuracy in time of war. There is no reason to believe that their aerial map knowledge of North America and other areas is as great as their knowledge of the territories over which rocket-firing trials have taken place. The USSR can never estimate in advance the extent to which electronic control and radio communications would be jammed by enemy counter-measures. They could of course in the next few years step up their long-range rocket production and give any surprise attack plan a high degree of probable success, but they cannot make the plan watertight. The consequences of a leak could be disastrous for the Kremlin and to a degree they cannot estimate.

Another uncertain element in Soviet strategy is the rôle to be played by her Communist allies. Some five or six years ago one might have envisaged a grand conception of Machiavellian manœuvring in the Kremlin, in which China, the Communist countries of Europe and eventually the countries of the Afro-Asian group would be knitted into a pattern of world domination by the Communist bloc. No doubt such a plan still exists with a number of Latin American countries fitting into the pattern; but it is fraught with difficulties of the kind that any grand alliance has always encountered. It is clear that China's policy in Tibet and on the Indian frontiers is not part of the grand Soviet design. In the middle 1950s the Kremlin stated that it was about to give atomic weapons to China and there were indications that the USSR was encouraging the formation of a Chinese Communist strategic bomber force. But for more than five years there has been no news of any further Chinese progress in this field and they have been five years of aeronautical plenty in which the Soviet Air Force could afford to hand over some of its long-range jet bombers to Pekin. And apart from China, to which if any of her allies shall the USSR hand over atomic weapons? Presumably Poland and Hungary by their separate actions in 1956 have forfeited much of the Soviet military trust. One wonders to what extent the Poles can be regarded as reliable allies of the Warsaw Treaty powers and whether they do not in fact constitute a serious security risk in any overall Soviet military planning. And if the Soviet zone of Germany were to be given atomic weapons would they be bound to use them politically and militarily in the Soviet interest? And now that there is every likelihood that African and Arab nations will achieve independence without a major Soviet intervention, does this not limit the rôle of the Kremlin in these countries' affairs to military and economic aid and to cautious cultural and political penetration?

One of the major defence reasons for the Kremlin to launch a surprise or pre-emptive attack is of course the Soviet inability to defend its home territory from the combined nuclear and thermonuclear assault of the United States and her NATO allies. Despite the production of many thousands of Soviet supersonic jet fighters and no doubt of many hundreds of ground-to-air guided missile batteries for local defence, the Commander-in-Chief of the Soviet Strategic Air Defence Command has no prospect currently of being certain of inflicting the minimum required loss rate of over fifty per cent against the 1,000-1,500 high flying jet bombers of the United States SAC strategic bomber command. The Western use of the new stand-off bombs with ranges of a hundred miles or more will certainly complicate the Soviet air defence problem. The technique of low-level flying attacks will give Soviet defences only patchwork warning even if their radar system is left unjammed. Western intermediate range and intercontinental ballistic missiles are likely to present an insoluble defensive problem to the USSR for many years to come. There have been no confident pronouncements on Soviet developments in the anti-missile missile field in a period when the Soviet military and political press is full of confidence about its technical achievements in most of the military spheres. General Pokrovsky, a leading Soviet rocket specialist, has said: "In principle a defensive anti-missile missile is possible, but of course such measures demand a complex development of technology and tactics." There is no reason to think that Soviet technology has made any more progress in this matter than the West, i.e. they may be developing long-range anti-missile radar and nuclear anti-missile missiles which may eventually be able to intercept a limited number of attacking missiles. But with each year that passes the advantage to the attacking missiles becomes greater if only because of the greater potential of submarine-launched missiles.

There is every reason to believe that the Soviet advantage in long-range missile attack will increase in the course of the next two or three years. Apart from a large numerical superiority over the West of perhaps three to one or more in intercontinental ballistic missiles, they are expected to have two or three hundred ready for operations in the next year or so. Even allowing for a firing or guidance error in the case of two out of three ICBM launchings, they may have enough missiles left to destroy or nullify most of the U.S.A. bomber and rocket bases. But in this dangerous period of Soviet strategic preponderance, in which a surprise thermo-nuclear attack on the West would seem to be the only and best logical military move by the Kremlin, a new factor will emerge which in itself may constitute a major new element in the Western deterrent. This is the new Missile Early Warning System which is expected to give up to fifteen to twenty minutes' warning of missile attack and therefore this amount of time for U.S. strategic bombers to scramble and take off from their bases. With this warning added to the proportion of U.S. strategic bombers airborne at all times and the Soviet uncertainty of attack against the Sixth and Seventh U.S. carrier fleets in the Mediterranean and the Pacific, there must be serious elements of doubt in the Kremlin about the prospects of any surprise attack on the U.S.A.

There is no doubt that the Soviet have certain permanent advantages over the West, particularly in military intelligence. While they probably know the location of all U.S. air and rocket bases both overseas and in the United States it is fairly certain that the West has a very imperfect knowledge of the exact location of Soviet air and rocket bases. There is reason to believe that the Soviet have a more highly developed passive defence organisation than the West and for other reasons have a better chance of surviving in a nuclear exchange of massive destruction. They can also allocate technical, personal and financial resources more easily than the West so that their current lead in long-range missiles can become permanent.

But although all these factors in the current military situation must give cause for concern in Washington and London they do not, in my view, constitute reasons for thinking that the Soviet Union will launch a surprise or pre-emptive attack, however logical it may seem from a military point of view. I do not believe that the Kremlin seriously thinks that Washington would in any foreseeable future launch a surprise attack on the USSR partly because the Pentagon cannot anticipate the extent to which such an attack would neutralise Soviet strategic air and rocket attack. Fear of such an attack would be the main reason for an initial Soviet attack.

Any summary of Soviet policy is bound to be a mixture of certainty and uncertainty. Certainty perhaps that NATO and the U.S.A. will not, in their own military interests, start an attack on the heartland of the USSR and that even if such an attack is launched, their power to retaliate should be sufficient to have a good chance of winning the war that follows with all its dire results. On the other hand if they themselves initiate mass thermo-nuclear attack, they must be uncertain of the strength of the Western thermo-nuclear reply and of the political consequences of what would follow such an attack. What would be the attitude of China to the USSR after the Kremlin had ordered the initial thermo-nuclear attack, but had survived the period of broken-back warfare which followed? Might not the Kremlin be surrendering to Pekin the strategic initiative in the Communist world which it now enjoys, and with it perhaps the leadership of the Communist world?

The gamble element in any thermo-nuclear war is clearly high for either side and the military record of the Kremlin since 1917 shows that it is not prepared to gamble in an uncertain situation. The earlier incidents against Japanese frontier forces in the 1930s, the limited contribution to the anti-Franco forces in the Spanish Civil War, the limitation of the Finnish campaign in the winter of 1939-40, the prudent if temporary alliance with Nazi Germany in the summer of 1939, the cautious moment of joining in the war against Japan in 1945 and the post-war incidents from Korea in 1950 to the Indian border in 1959 have all underlined the Soviet unwillingness to risk launching a major war on her own account.

However logical it may be for the Soviet authorities to initiate a surprise attack on the U.S.A. and Britain in the next year or so the elements of a major gamble will always be there, and will be increased within the next few years, despite the potential growth of Soviet preponderance in strategic attack. The development of the new Western Missile Early Warning System in itself can, as we have seen, take much of the surprise effect out of a surprise attack. The fact that the Kremlin will have to plan a mass simultaneous rocket and bomber attack would give radar maximum opportunities of providing effective early warning. It is doubtful if the Soviet authorities can make any realistic assessment of the damage potential of the combined retaliatory Western nuclear and thermo-nuclear forces even if they had full knowledge of all the bombs, missiles and aircraft involved. This would mean assessing in advance the potential of untried radar equipment on both sides. All past war experience goes to show that such assessments are not feasible, that is with any degree of certainty. A single piece of jamming equipment which chances on a frequency at the correct distance from the radar early warning position can change the entire defence picture. And with every year that passes the U.S. potential in hardened land-based ICBM rocket sites increases as does the potential of Polaris submarine-launched missiles and mobile solid fuel Minuteman rockets which can be mounted on a mobile base such as a train or a ship.

The Soviet dilemma, assuming that they wanted to launch a surprise attack, is that they have not yet the long-range rocket forces to ensure its total success and that by the time they have the resources there is likely to be much less certainty of success because of U.S. long-range radar at first and hardened sites and greater mobility of bases later. One wonders, however, why the Western powers have not been able to disperse their bomber and rocket forces even more. For instance in an age in which mobility and dispersal are more important than ever for NATO and U.S.A. it is hard to understand why Canadian real estate is used primarily for housing defensive fighters and missiles and not for Western bombers and long-range missiles. The ability to use Canada to the full for NATO bomber and rocket sites could in itself present the Kremlin with such new planning headaches that these would constitute the final guarantee of world peace by making the Soviet press on more realistically with its disarmament proposals. This would be particularly true if the details of the sites were kept out of the Western press and not presented to Soviet intelligence as they so frequently are by over-zealous British and U.S. journalists. There was documentary evidence to show that British and U.S. newspapers and magazines were the most important and cheapest sources of military intelligence in the Second World War. It would be an odd historical commentary if an archivist found in fifty years' time that the democratic desire for freedom of the press had been a major cause of the downfall of Western democracy.

What would be interesting to know is the extent to which the Western underestimate of Soviet potential strategic air and rocket power led to the present situation in which the Western deterrent has become less credible and less effective than it was five years ago. Certainly it is hard to avoid the conclusion that there was a serious underestimate of Soviet long-range military potential in the highest quarters on both sides of the Atlantic, in the immediate post-war period. The U.S. Defence Secretary put the 1950 Soviet aircraft production figure at 600 planes a month and the British Defence Minister put the annual figure at about 10,000 planes in the summer of 1951. It is hard to see how the Soviet Air Force could maintain an operational strength of fifteen to twenty thousand aircraft and supply planes to more than half a dozen prospective allies on such production figures. Hanson Baldwin in *The Price of Power* estimated that the Soviet would probably have an atomic bomb between 1950 and 1957 and that between 1960 and 1970 the Soviet pile might be numbered in three or even four figures. One of the most striking underestimates came from a Russian airman and engineer who has in so many other ways shown himself to be a man of extraordinary military vision. In *Air Power—Key to Survival* Alexander Severski wrote of the USSR (page 15): "No unfree society can hope to match us on the highest levels of technology skills and individual initiative expressed in their ultimate in modern air power." And he went on later (page 26) to add: "In the case of Soviet Russia, however, I am convinced that our margins of social and technological strength are so great that the policy of deterrence can be valid for the foreseeable future." It is probably invidious to single out such quotations, for no Western writer on Soviet military affairs has been free of such errors of underestimation in the past. To publish assessments of Soviet air power is to realise that, as has been said, there are no experts on the USSR, there are only varying degrees of ignorance.

But if there has been underestimation in the 1940s and early 1950s, there has surely been the danger of overestimation since the Kremlin announced that it had successfully test-fired an intercontinental ballistic in the summer of 1957. When the news broke, the Western press, including the serious newspapers, gave vent to a burst of alarm and pessimism which suggested that the entire Western strategy of deterrence had already failed and that the political and strategic initiative had passed irrevocably to the Soviet Union. And yet it was a liquid-fuel missile which was tested. These according to Western scientists are very susceptible to misfirings and other inaccuracies, and the USSR at the moment still relies on liquid-fuel rockets for long-range rocket attack.

If one looks back on the past three years since the USSR successfully launched its first earth satellite a period in which we have had ample proof of extraordinary further progress in Soviet rocket engines and guidance systems, can one say that the foreign policy of the USSR has been more aggressive and confident than for instance before the summer and autumn of 1957? Have Soviet dealings with the Shah of Persia or Colonel Nasser indicated any greater sense of strategic *insouciance* in the Kremlin? Was Mr. Khrushchev really flying a kite on Berlin or uttering a serious military threat to the West in 1959? At the time it seemed a threat but when the deadline of May was reached his statement seemed more like a speculative speech than a prelude to aggression.

Introduction

In the chapters that follow it is hoped that a reasonable balance has been maintained between optimism and pessimism and that it emerges clearly that disarmament is a logical military policy for the Soviet Union despite any strategic or tactical advantages she may enjoy in defence matters now and in the future. In considering the future of Soviet air and rocket power in the final chapter, there is inevitably some return to themes and views discussed in earlier chapters of the book. It is hoped that the repetition is not too tedious for the reader and that critics will be as helpful as on previous occasions in pointing out errors of fact and thought.

There have been many new winds both political and military blowing through the Kremlin corridors since Stalin's death. Like all winds, they are not easily susceptible to accurate forecasting of speed and direction. We have for many years now been saying that the USSR has held the strategic initiative. I believe, however, that Western military policy has presented the Kremlin with a number of military problems and compelled it to readjust its planning in the last five or ten years. Before 1955 I doubt if it was intended to base Soviet military planning on the mass use of tactical and strategic atomic weapons, but Mr. Khrushchev's statement on defence policy to the Supreme Soviet early in 1960 made it fairly clear that that is now the case. In fact when one looks back on the military history of the 1950s one may find a great similarity between the weapons and policies of the West and of the Soviet Union. And in the identity of long-term planning and weapon development one finds that the Kremlin ideas have evolved some years after those of the West in such areas as the development of a strategic bomber force, of a long-range military air transport force, of the large-scale use of tactical atomic artillery and missiles and in the various uses of radar.

It would be surprising if the military text books of the USSR to be published in the next few years differed essentially from those published in the West, though they will no doubt be liberally sprinkled with anti-Western dialectics and have a greater political content. Certainly one has had greater access to Soviet military staff thinking in the past few years and this is due largely to the publication of three new books. The first was *Soviet Strategy in the Nuclear Age* by Dr. Raymond Garthof (Praeger, New York, 1958), the second *War and the Soviet Union* by Dr H. S. Dinerstein (Atlantic Books, London, 1959) and the last *Science and Technology in*

Contemporary War by General G. I. Pokrovsky (Praeger, New York, 1959). Of these three books the last is perhaps the most important, for the material in it must have been widely read by members of the Soviet air and rocket forces in the past few years. Pokrovsky, now aged about 60, is a general in the engineering technical services of the Soviet Army, a member of the staff of the Zhukovski Air Engineering Academy and has personally played a major part, as a nuclear physicist, in the development of Soviet nuclear weapons, long-range missiles and earth-satellite sputniks. A distillation of his views on contemporary warfare is therefore of some importance. Equally one should perhaps point out that the views expressed in this book are mainly the result of studying the trends and data of Soviet air and rocket weapons rather than of pronouncements by Soviet leaders, military or political.

Unfortunately Pokrovsky's writings are tinged with the usual warpings of his ideological preconceptions; he states for instance that "Soviet military science has a fundamental advantage over the military science of the capitalist countries" but between the recurring bursts of political diatribe he provides many examples of authoritative insight into contemporary Soviet military thinking which elucidate Soviet weapon and other developments. One of the key passages in his writing reflects the revolution in Soviet air thinking in the 1950s. It runs as follows (page 69): "It would be a great error to under estimate the significance of strategic weapons in a possible future war. It is a strategic weapons system which can most of all lead to the unexpected in military affairs. Strategic armament can completely change the whole character of warfare converting the entire territory of even very large countries into arenas of combat and even extending the area of combat in the near reaches of outer space." Such a pronouncement by a Soviet general before 1950 would have been impossible. General Pokrovsky goes on to say (page 71) that "long-range rockets should be put in first place in a list of the forms of military technology". He makes the further point that it is difficult for the enemy to detect the rocket sites which can be widely dispersed and camouflaged. This is of course much truer of Soviet rockets than of Western long-range rockets and this difference is a crucial factor in assessing the East and West deterrents. One surprising Pokrovsky remark is (page 74) that "the long-range rocket launched by the enemy by surprise will be detected by radar only several tens of seconds before the moment of explosion"; and he adds, "in practice the long-range rocket ensures surprise attack on distant targets". It would seem that the possibility of a long-range early warning of up to twenty minutes enabling retaliatory bombers to get off the ground was not a part of his thinking in this context. Interestingly enough, Pokrovsky, writing in 1957, said: "Long-range and intercontinental rockets cannot completely replace piloted aircraft" (page 179), and in gauging rocket accuracy he said: "One can suppose that the landing point of intercontinental rockets does not deviate from the target point by more than ten to twenty kilometres." This is a much more modest claim to accuracy than has been put out by Soviet politicians and military leaders. However, the tests of Soviet rockets in the last year or so suggest that greater accuracy has now been achieved; but even if this is so, Pokrovsky points to another long-range rocket limitation. "It is necessary to know in advance the location on the map of the target and the map must be sufficiently accurate to pinpoint the actual location of the target on the earth." It is when dealing with radar and electronics that Pokrovsky expresses his strongest doubts and says (page 101): "One must not exaggerate the advantages of these new [rocket] weapons. The more automatic any procedure becomes, the easier it becomes for the enemy to jam that procedure. The jamming of self-seeking guided missiles can be accomplished both optically and by radar. It is more difficult to jam the autonomous guided missiles in which guidance mechanisms are installed to assure flight by a predetermined programme. But these missiles can be used only when precise advance knowledge of all the conditions of the combat situation is present. Manœuvre can easily fool such automatic weapons."

It seemed worth quoting extensively from General Pokrovsky's writing if only to show the difference between the confident bombast of pronouncements by Mr. Khrushchev or his Defence Minister and the more realistic assessments of leading Soviet rocket specialists. Even the published text of Pokrovsky is likely to be more confident about the new rocket weapons than some of his inner councils in the Soviet engineering academies.

Pokrovsky seems to be a little more sanguine about the future of manned aircraft than other Soviet military and political leaders including Mr. Khrushchev. Reference has already been made to his appreciation of the value of the manned long-range bomber. He also sees a future for tactical bomber and fighter planes. Missiles, he points out, can have their guidance knocked out "leaving the operator powerless and passive. Aviation can in the future destroy roads, canals, etc. . . ." Pokrovsky, however, warns against the dangers and complications of high-speed low-level attacks. "It will not be difficult to erect barriers against such types of aircraft." He sees increased rôles for transport aircraft and helicopters in the future with special importance attached to vertical take-off planes, both piston-engined and jet being in general use for local communications work. But he also sees a future for vertical take-off combat aircraft, e.g. to avoid surprise attacks on air bases. He writes: "Supersonic aviation without airfields can be based in individual natural shelters; in woods, in mountain ravines, and in other terrain difficult of access and suitable for camouflage . . . the rôle of fighters with rocket boosters which can be employed without the need for conventional airfields will increase."

One has quoted Pokrovsky to show not only the way Soviet air and rocket strategy is tending to go but also to show some of his doubts. He begins his book with the sentence, "Contemporary warfare is a very complicated process", and goes on later to talk about the importance of the theory of probability in war. "The theory of probability has long been employed for artillery and bomber aviation. Now it is beginning to be widely introduced into military affairs as a whole."

It is the element of doubt added to the holocaust of destruction of which both sides are capable which leads one to the conclusion that a position of strategic stalemate will persist through the 1960s and so give time for the hesitant and uncertain disarmament councils. Indeed the greatest comfort one derives from writing about Russian air power is the feeling that it will never be exercised except in harmless manœuvres or in peaceful competition on the passenger and freight air lines of the future.

CHAPTER I

RUSSIAN CIVIL WAR TO SPANISH CIVIL WAR

THE aviation inheritance of the first Soviet government of 1917 accruing from the former Russian Imperial government was scanty in the extreme. It was on December 20th, 1917, that the new Soviet rulers first promulgated a decree setting up both aircraft and airship formations. The airship formations took virtually no part in the Civil War of the next few years and Soviet aircraft units were, for a long time, understrength, ill-manned and fitted with obsolescent equipment. The year 1917 was scarcely the ideal moment to create a new air force in Russia. It was a period of violent revolutionary chaos. In the current disorders, some of the limited number of aircraft factories were being wrecked, pillaged or abandoned. Technicians and engineers were fleeing to country districts in search of food or to other countries in search of political sanctuary. But despite widespread disorder and endemic shortages of aircraft fuel, spare parts, aircraft maintenance facilities, air technicians, training schools, trained personnel—in fact everything which goes to make the nucleus of an air force—the Soviet authorities, with typical revolutionary fervour, disdained the handicaps and made do with what they had. Their 1917 aircraft assets consisted of some 300 ill-assorted machines in varying states of serviceability. They were mainly French Caudron and Morane planes, British Napier machines, some Italian Ansaldos and finally some Russian-built Grigorovich seaplanes and the Ilya Mouremetz heavy bombers. The last was one of the first four-engined bombers in aviation history; it was at first powered by four French built 600 h.p. engines, but by 1917, when it had been used successfully by Russian naval air squadrons against naval and army targets on the Baltic front, the Mouremetz heavy bomber was fitted with four similar Russian-built power units of French conception.

When the Civil War broke out in the autumn of 1917, the Soviet authorities acted swiftly. The Central Military Revolutionary

Council quickly created an air staff responsible for the training, organisation and operations of all airship and aircraft units and subordinated them entirely to the combat requirements of the Red Army. The new Soviet air headquarters in Moscow, formed in 1917, established in each of the main combat theatres a "Field Administration of Aviation and Airships", which, for the next three or four years, was responsible, under Red Army aegis and supervision, for the conduct of air operations against the White Russian and foreign interventionist forces. The early handicaps were gradually overcome by a mixture of zeal, repression, threats, executions, patriotism and Bolshevik fervour. By 1919, the new Soviet Air Force had a statuory strength of some 350 aircraft, though less than half this total were simultaneously fit to fly and then for not more than one or two operations in succession. The bulk of the air units were to remain below target strength for most of the revolutionary war. Fighter and bomber squadrons, which should have had an operational strength of nine or ten machines had to be content with a fighting strength of some half a dozen machines, sometimes less. The twenty float-plane squadrons had particular difficulties in geting replacement machines and had to break up aircraft in order to get spare parts. There was a nominal long-range bomber air division of Mouremetz four-engined planes which could rarely put more than a dozen machines into the air. The five divisions of airships rarely went into action.

But despite all the current problems of personnel and supply, two or three new operational air bases were soon built and the existing flying training schools equipped and manned. In addition, some half-dozen air supply depots and maintenance and repair centres were established. But it was the shortage of skilled and experienced flying personnel which handicapped the new Soviet Air Force. So many of the best Tsarist pilots had come from the suspect middle or upper classes. Those who were not shot made their escape, some of them like the air ace Kazakov to fight in the Civil War against the new Communist Air Force of the USSR: others, like Alexander Severski, went to the U.S.A. But a fair number of experienced flying personnel stayed with the newly born Soviet Air Force. Meantime, the Revolutionary Council adopted a ruthless method of aircrew selection. They simply hand-picked young Communists for their physique and ordered them to fly. No figures are available for training crash rates during this early formative period, but they must have been formidable. There was an acute shortage of flying instructors, airfield conditions at flying training schools were deplorable, aircraft servicing standards were low and repairs were often carried out by an unskilled aerodrome worker improvising with wire and string. Over half the surviving pilots of the 1917-1920 period were Communist Party members and they achieved heroic if desperate flying exploits in the Soviet air operations of the Civil War.

These were almost entirely tactical in nature, consisting of close fighter, bomber and reconnaissance operations on behalf of Red Army land battles. To ensure good liaison with ground forces, the under-strength air squadrons and air divisions were usually attached for fixed periods to Red Army divisions. The general scale of air operations carried out by the new Soviet Air Force was extremely restricted and they rarely achieved even temporary local air superiority. They were frequently heavily outnumbered. On the Archangel front, for instance, the White Army had the support of up to one hundred planes while the local Soviet air units could rarely muster even a dozen obsolescent serviceable machines to support the Red Army forces. Other Soviet air fronts were even worse off. On the Astrakhan front there was a single air squadron (No. 47) which often had a serviceable strength of only one or two aircraft. The lack of air support on this front was having such an adverse effect on Red Army ground troops that Kirov, the local representative of the Military Revolutionary Council, made stern representations in Moscow for greater air support. The net effect was not to increase the number of Soviet air squadrons on the front but simply to bring the existing squadron up to its statutory strength of about ten machines. Budenny's entire crack cavalry army could only command the air support of two squadrons and these were under strength, comprising some fifteen aircraft and sometimes less than a dozen pilots.

Nevertheless, Soviet Moscow staged its first public air display as early as May 1st, 1918, and used the occasion to extol the virtues of Soviet aviation and aviators and to denounce capitalists and imperialists in the most virulent terms. The pattern of commentary for these annual flying displays were thus set for the next forty years or so. To Western ears these annual diatribes against the West are dreary, inexplicable and unjustified. For the Soviet, it is a vital part of the

Marxist myth which has to be reiterated in one form or another on most public occasions, even in the dangerous age of atomic weapons.

In the main, the new Soviet air units did not have a very impressive operational record in the first two years of the Civil War from 1917 to 1919. The puny resources had to be dissipated on too many battle fronts. The Soviet Air Force was, in this period, making a token effort at air operations everywhere without being effective anywhere. At the end of 1919, Red Army leaders met to consider their current air policy. There were strong Army protests at the spreading of their meagre air resources over such a wide area and the general view was that a more concentrated effort in selected areas should be made. A new directive was issued on how to deploy and operate the air units and this had the effect of denuding the secondary unimportant war fronts of their previous penny packet air support and concentrating the main effort, such as it was, on the important fronts.

It is at least of passing interest that even at this early stage the Soviet air doctrine was flexible. It was subject to alteration as the result of the common-sense dictates of Soviet military resources. This has remained a cardinal feature of Soviet military air policy over the last forty years. While the pronouncements of Stalin, Khrushchev or other Soviet military leaders may have seemed hide-bound by Marxist or Communist, particularly Stalinist, propaganda, Soviet military practice, production policies and training schemes have been realistically adapted to the practicalities of the current military situation. The new air directive of 1919 to concentrate the effort on the important battle fronts was implemented at once. On the Baltic front as many as fifty aircraft were assembled and operated in support of land and naval forces which were meeting the threat to the key Communist town of Petrograd (Leningrad) at the end of 1919. A twenty-five kilogramme fragmentation bomb was the heaviest used in these operations and formations of more than five or six aircraft were rarely engaged. All this may seem pigmy by World War II or current standards, but it was quite a respectable concentration of air power at the time. A further example of the new concentration policy was the use of some 20-30 aircraft to meet the threat of General Marmontov's cavalry which had penetrated Red Army positions on the southern front in the Ukraine and was threatening to outflank or perhaps encircle them. In this battle Soviet air units persisted with the technique of lower level attack against enemy ground forces for the first time and were pleased with the results. The technique was repeated successfully in a subsequent air action against General Wrangel's cavalry forces. Mainly as the result of these two local successes and earlier experience, two air principles were established for the new Soviet Air Force. The first was the creation of specialist Sturmovik ground attack air units, with a strength of up to 40 planes, which foreshadowed the successful Ilyushin (IL 2 and IL 10) Sturmovik air regiments of World War II. The second was the recognition of the importance of local air superiority, and so of adequate fighter escort to prevent the enemy from breaking up the air attacks; and so the bulk of Soviet fighter aircraft resources were used in support of the new Sturmovik tactical bomber operations. And thereafter, right up to the 1950s, the élite of Soviet fighter units were to be used in this general rôle. Partly as a result of these 1919 directives and experiences, the sorties of the Soviet Air Force in the second half of the Civil War from 1919 to 1921 were much more effective than in the first half. During the latter period, some 10,000 sorties are estimated to have been carried out. In the earlier period only 2-3,000 sorties were made.

The Soviet government was, from the very start, conscious of the importance of keeping pace technically with aircraft research and development both in Western countries and in Japan. It gave full backing to the new Soviet Air Force from the beginning. As early as 1918, Zhukovski, the doyen of Soviet air engineers, established the Central Institute of Aerodynamics and Hydrodynamics (TSAGI), which to this day controls to a considerable extent experimental development and testing of all Soviet aircraft both military and civil, though local initiative has been always encouraged. By the time of his death in 1921, Zhukovski had assembled at the new Institute a group of enthusiastic aeronautical engineers and designers, many of whom were later to be leading figures in the Soviet aircraft industry in the jet atomic age: men such as Tupolev, and the engine designers Klimov and Mikulin. Other leading figures of the 1920s, some of whom were liquidated in the purges of the 1930s, included Kalinin, one of the earliest designers of swept-wing aircraft, Grigorovich, Chaplygin, Alexandrov, Yuriev, Bratukhin and Mil. The latter was to be a pioneer Soviet designer in the helicopter field. In his brief spell of service for the new Soviet government, Zhukovski did much to lay the foundations of technical air training and aircraft design in the USSR. He set up the first air technical school in Moscow in 1919 called the "Institute of Red Air Fleet Engineers" which was renamed the "Zhukovski Military Air Academy" when he died. Most senior technical Soviet Air Force officers of the past forty years have taken an instruction course at this academy, including such leading present air personalities as jet aircraft designers Sergei Ilyushin, Alexander Yakovlev and Artem Mikoyan. The new technical air college gave the engineer officers of the new Soviet Air Force who graduated from it a higher status than their opposite numbers in other air forces. This was not surprising in a new régime which was to make a religion of technical progress right from the early years. But in the 1920s progress was slow. A November 1919 decree issued by the Soviet Council of Labour and Defence had called up all engineers and technicians with any experience in the aircraft industry. The response did little to improve the acute shortage of skilled personnel which led the Soviet to the early necessity of importing them from France, Germany or indeed anywhere abroad.

When the Civil War ended in 1921, the Soviet authorities were faced with the huge problem of trying to make a largely peasant population both air and technically minded. The bulk of the aircraft and aero-engines used in the Civil War had been of French and Italian design. TSAGI had to start virtually from scratch and with limited resources. But perhaps this was not as serious as it sounds; in the 1920s military and civil aviation were still in their infancy and pleading for general recognition. By 1920 and 1921 the first Soviet-built 200 h.p. aero-engines were being produced at a factory near Moscow with some help from French engineers. Indeed outside the factory for many years could be seen a brass plate marked "Moteurs Muscovites". Junkers also helped to build near Moscow an aircraft factory called the Fili plant which by 1922-23 was turning out the ANT 1 aircraft designed by Tupolev. German resources were also used to start the first passenger air service from Moscow to Königsberg in 1922. Some of the post-civil war aircraft came from Junkers factories in Sweden, others were Savoia float-planes from Italy; there were British reconnaissance machines from Vickers and de Havilland, and Fokker planes from Holland. It was Lenin's policy to turn to the capitalist world for technical help and nowhere was this policy carried out more intensively than in the Soviet Air Force.

If there was extensive foreign help, there was also a great public-spirited drive to make the USSR an air-minded nation with an eye on the future. The Communist Party and the Komsomol were naturally in the van of the new campaign propaganda. Much of the publicity intended to help make the new Russia air-minded was carried out by an organisation founded in 1923 called the "Friends of the Air Fleet Association" (ODVF). It launched a nation-wide savings campaign which, in its first drive, produced enough roubles to build about a hundred aircraft. It attracted about a million young Russians to the 5,000 branches of the ODVF which were set up all over the USSR. Workers were encouraged to give up a day's pay to the new organisation and no doubt the encouragement was often pressurised. However, what was important in the ODVF organisation was not the thousands of millions of paper roubles and the plastering of walls with aeronautical expressions of enthusiasm. It was part, and an important part, of the initial Bolshevik enthusiasm, devotion and *esprit de corps* which later on helped to sustain the Soviet Air Force through the dark patches of its early career both before and during World War II. The Russian man or woman has a greater than normal capacity for hard work, loyalty and enthusiasm. The Slav compound of kindness and cruelty, bravery, ruthlessness and virtuosity found in the Russian character is obedient to a single unifying idea such as devotion to air power or science or space travel. Such press slogans in the 1920s as "Fight on for ten more planes" may seem absurd to the non-Russian observer, but they were examples of the normal journalism of the USSR in the 1920s and they helped to achieve results.

One of particular importance was the promotion of the new Osoaviakhim society (the letters stand for the Association for the Promotion of Defence, Aviation and Chemical Warfare). Despite the last part of the title, the organisation is not known to have included chemical warfare in its activities. Instead of the usual English rendering, perhaps "scientific warfare" is a more appropriate translation. Osoaviakhim in fact was a para-military organisation for Soviet youth which took over the work of the ODVF organisation and expanded it: it co-ordinated the civil defence training of the population, promoted parachuting, flying and glider training, provided initial technical training for wireless operators and machine gunners and generally did anything to create a national appetite for aviation. The gliding schools were especially important. Whole teams of glider designers were urged to get down to the business of providing sail-planes for the Osoaviakhim youth, including leading designers such as Antonov, Yemelianov, Gorshkov, Grigovski and Sheremetiev. Well-equipped glider clubs were established near Moscow, Stalingrad, Tula, Leningrad, Engels, Kirovabad, Kiev, Tashkent and Tomsk, in fact all over the Soviet Union.

Until the 1930s, the Soviet aircraft industry continued to rely heavily on foreign planes, aero-engines and ideas. For the first years after the civil war, Soviet designers concentrated on producing small aircraft which could be used for training, communications and commercial aviation. By 1926, although Tupolev had designed the ANT 1, 2 and 3, they were only two or three-seater machines. By 1928 he had designed a twin-engined ANT 4 suitable for civil air line operations, a prototype four-engined machine, and by 1930, the ANT 9, a three-engined machine similar to the Junkers 52.

At the end of 1928, the aircraft production scheme for the First Five Year Plan was formulated and the announced intention was that the Soviet aircraft industry should "rid itself finally of foreign dependence". In the event, aircraft production in the USSR was to depend heavily on foreign designs and technical information and prototypes from outside the Soviet Union for the next twenty to twenty-five years. However, in the First Five Year Plan period (1928-32) the Soviet need to rely on foreign-made aircraft was substantially reduced and the shape and characteristics of the first generation of Soviet-built machines began to emerge. In the single-engined fighter category, three aircraft were put into production, a single seater biplane designed by Tupolev called the I 4, a more advanced fighter the I 5, designed jointly by Polikarpov and Grigorovich, and finally a Grigorovich fighter, of which only a few were produced, but which was one of the first machines in military history to be fitted with a cannon. Then there was a two-seater fighter category, intended mainly for the Sturmovik ground attack units. In this class, the First Five Year Plan saw three types of aircraft in production in Soviet aircraft factories, the DI 2, 3 and 4, all designed by Polikarpov. However, neither Polikarpov nor Grigorovich (the latter also designed some float-plane fighters during the First Five Year Plan) were to have any long-term success as designers of Soviet fighter bombers and Sturmovik aircraft. In the period just before

World War II, their work in this field was taken over by a more successful group of engineers including Lavochkin, Ilyushin, Gurevich, Mikoyan and Yakovlev, who were to be the chief Soviet designers in these categories in World War II, and also in the post war jet age.

Perhaps the most interesting aspect of the First Five Year Aircraft Production Plan was the attempt to produce a range of so-called heavy bombers. Tupolev and his colleagues at TSAGI clearly believed in the future of both heavy bomber and transport aircraft in this period, but there was apparently the same dichotomy in military thinking on this subject in the USSR as there was in other countries at the time. In view of the shortage of skilled workers and raw materials, including aluminium, and the domination of Soviet strategy by Red Army thinking, it is not surprising that Tupolev and his colleagues had to be content with modest beginnings in this field of aircraft design. Although they designed two aircraft, the TB 1 and TB 2, which were nominally heavy bombers (T standing for *tyazhely*, the Russian for heavy) they were in fact twin-engined machines which would have been more correctly labelled medium bombers. Tupolev, however, must have had considerable influence in Soviet aircraft policy councils, for at the end of the First Five Year Plan, he designed and developed a four-engined machine, the ANT 6, which became the TB 3 heavy bomber. By the middle 1930s, it was to be the only long-range four-engined bomber to go into service in large numbers in any air force between the two world wars. There are a number of reasons why the Soviet Air Force did not maintain its initial lead in developing heavy bombers, for instance the strong rival claims of the parachute and airborne formations for long-range transport aircraft and the relative failure of Soviet bomber operations in the war in Spain. But there was also the purging of air and military leaders, such as Tuchachevsky and Khripin who were devotees of the importance of strategic bombing in modern war.

The First Five Year Plan certainly failed to lay the ground for the air transport requirements of the 1930s. In February 1923, the Soviet Council of Labour and Defence had passed a resolution calling for the creation of a network of civil air lines to carry passengers and freight. But despite the obvious importance of the new Aeroflot civil aviation organisation in a vast country like the USSR, equipped with such limited rail and road networks, for at least the

first two decades of Soviet history, the development of native designed transport aircraft failed to keep pace with those in service in civil air lines and military air forces outside the USSR. TSAGI did not neglect this field of design, but its efforts were dogged by a series of failures and disappointments in both the First and Second Five Year Plans. A German Junkers three-engined blue-print was developed and became the ANT 9, Tupolev also designed the single-engined ANT 25 and also a five-engined 34-seater prototype which never went into series production. However, there were in the early 1930s few air transport squadrons in any air force in the world and so the relative Soviet failure was not serious at this stage of its air transport development.

The First Five Year Plan belonged to a period of aircraft design in which float-planes and flying-boats figured prominently in military and naval air thinking, especially in Germany and Japan. It is perhaps surprising that the Kremlin should have shared in this thinking, for the Soviet Navy exercised no great influence at the time and tactical support of the Red Army was the basis of Kremlin air policy. However, the prospect of air defence and reconnaissance operations over the Baltic and the Black Sea was there and the fine record of the Russian Naval Air Arm in World War I may also have been a factor. Whatever the causes, TSAGI sponsored the design and production of a number of float-planes and flying-boats in the 1928-32 period. In the fighter category, the MI 4 was to have a short-lived career and within a few years the Naval Air Arm was re-equipped with land-based fighters and this tradition has remained to the present day. But the long-range reconnaissance flying-boats designed in this period (the MDR 1 and 2) were to remain in service for many years to come; the USSR also built a Grigorovich twin-engined float-plane equipped with two 500 h.p. engines, giving a top speed of about 150 m.p.h. and a range of about 750 miles, i.e. a performance roughly equivalent to many naval aircraft in this category in other air forces of the early 1930s.

The two chief Soviet aero-engine designers of this period were Mikulin and Svetsov, both of whom survived the purges of the late 1930s. In fact, both were amongst the most active and prominent Soviet engine designers twenty years later in the jet-atomic age. They began work on the five cylinder 100 h.p. M 11 radial engine of French inspiration, but by the end of the First Five Year Plan period

they were working on 500–600 h.p. engines, including the liquid-cooled AM 34, which was later boosted to 750 h.p., the AM 17 which was to give 600 h.p. and the M 22, a twelve-cylinder engine of German BMW inspiration. These were the successes: but the First Five Year Plan inevitably also had failures. Bessonov, one of TSAGI's leading aero-engine designers of this period, made three successive false starts which did not go beyond the blue-print stage. Other almost inevitable aircraft failures of this period were the Soviet autogyros. Two of them were successfully test flown in the 1930–32 period. The leading designer in this Soviet group was Mikhail Mil, nowadays the most prominent of Soviet helicopter designers. Working with him on autogyro design in the early 1930s were such prominent present-day helicopter personalities as Kamov, Bratukhin and Kupfer. But like other autogyro and helicopter designers in Britain, Spain and Germany they were ahead of their time. Can any aeronautical historian satisfactorily explain the almost universal prejudice in air circles against vertical take-off aircraft in the 1930s and 1940s? The Korean War marked the first occasion on which helicopters were used on a large scale. The 1960s will certainly see them used increasingly for transport, reconnaissance and anti-submarine work.

The Soviet First Five Year Plan for aircraft production made a useful start in the reconnaissance aircraft category. Tupolev's ANT 2 biplane was manufactured in a specially made version for patrol and observation duties and was designated the R 3 (R stands for *razvedchik*, the Russian word for reconnaissance) and Polikarpov designed the specialist biplane R 5, which was to be used on reconnaissance missions in the coming Civil War in Spain. Tupolev also developed a reconnaissance version of a twin-engined bomber which was designated R 6. The surprising thing is that no specialist reconnaissance aircraft for artillery spotting was devised for an army which rated its artillery arm so highly. Once again this was true not only in the Soviet Air Force but in many other contemporary air forces. Spotting for artillery was one of the first major functions of military air power in the 1914–18 War. One would have imagined that specialist aircraft spotting would have been a major preoccupation of subsequent military aircraft designers. But the fighter, bomber and transport machines seem to have pushed the specialist reconnaissance aircraft lower in the production priority lists both in the USSR and in other countries.

Looking back at the period of the First Five Year Plan for Soviet aircraft production gives one an impression of success and drive achieved in the face of many handicaps. Stalin's claim made in 1933 that the Soviet Union now possessed its own native aircraft industry could be substantiated in the main. However, VIAM, the Soviet Institute for Research on Aircraft Materials, was still making extensive use of foreign sources and TIAM, the Institute for Aero-Engine Development, was working intensively on the adaptation of foreign designs to meet Soviet requirements. There was still a general shortage of skilled workers, technicians, machine tools and raw materials. In most cases the target production figures were not achieved. The twin-engined TB 1 and TB 2 medium bombers for instance were due to reach an annual output figure of 500–600 a year in the early 1930s, but less than half this total were in fact produced. It was all very well to put in a corps of shock Stakhanovite workers at the airframe plant at Zaporozhe or at the aero-engine plant at Taganrog. Most of them were not skilled workers and the net result was that the assembly work was often so crudely done that many aircraft failed to get off the ground in their initial factory flight tests; engines failed frequently and had to be scrapped or replaced all too often. But the broad achievements of the First Five Year Plan were quite impressive. Between 1,000 and 1,500 aircraft were produced annually at the beginning, and the total had reached nearly 2,000 a year at the end of 1932. The appointment of aeronautically illiterate party members to key posts in the aircraft industry must have hampered efficiency, but it also added to the dynamics of the Soviet aircraft factory workers at Leningrad, Moscow, Rybinsk, Voronezh, Zaporozhe, Taganrog and elsewhere. At this stage it did not matter greatly if half the finished products had to be wasted and rejected or that Soviet bombers were virtually copies of Junkers designs, the fighters of Heinkel design, the reconnaissance aircraft inspired by British de Havilland machines and the flying-boats based on Italian Savoia blue-prints. The important thing was that the USSR had by early 1933 a front-line strength of between 1,000 and 1,250 aircraft organised into some twenty air brigades.

There were reconnaissance and fighter units based at Moscow, Smolensk, Sevastopol in the Crimea, Kiev, and Kazan and bomber units near Moscow, Leningrad, Rostov, Sevastopol, Kazan and Tashkent. The naval air units were located near Leningrad and

Sevastapol, while Rostov was an important operational training base for naval air squadrons.

By now most of the aircraft had been made in Soviet factories. There was no longer any need to import aluminium from Germany, for Soviet scientists had discovered a native substitute for it called Kolchougalumin. When Hitler came to power, there was less need to use German technicians and flying instructors as the USSR had done in the late 1920s.

The First Five Year Plan for aircraft production had laid reasonably good foundations for the more advanced Second Five Year Plan period (1933–37). The international situation was now deteriorating fast. Soon Japan was to invade Manchuria, and the Anti-Communist Pact between Germany, Italy and Japan was to be signed. Germany was to reoccupy the Rhineland and to go on to occupy Austria and Czechoslovakia. How did Soviet aircraft production and technical standards measure up to the new international situation during the middle 1930s? In view of the increasing threat from Japan it was natural that the Second Five Year Plan should have included a scheme to develop a self-supporting Far Eastern Air Force. It was during this period that new aircraft factories were set up beyond the Urals at Irkutsk, Tomsk, Khabarovsk and Komsomolsk. There was also an expansion in air training facilities in the eastern areas of the USSR. New schools for both military and civil flying were set up near Samarkand, Khabarovsk, Kupino, Novoskibirsk, Frunze, Tashkent, Irkutsk, Abakan, Barnaul, Tomsk and Sverdlovsk. Instead of a few scattered air units, it was now possible to plan the creation of a Far Eastern Air Army of a strength of up to 1,000 aircraft, with integrated air brigades of bomber, fighter and reconnaissance units. The overall front-line strength of the Soviet Air Force was roughly doubled in the period of the Second Five Year Plan from 1,000-1,250 to over 2,500 planes, divided into some forty to fifty air brigades each of about 50 machines, and stationed in the main fifteen military districts of the USSR.

What of the quality of aircraft and the training standards of the crews which, within a year or so, were to have their baptism of fire against German and Italian squadrons in the Spanish Civil War and against the Japanese Air Force? The native-built Soviet fighter planes designed and produced in the Second Five Year Plan period did not measure up to the requirements of the late 1930s and the initial operations of the Second World War. Polikarpov's I 15 was first flown in October 1933 and was powered by a 700 h.p. engine but its maximum speed was only about 225 m.p.h. It had an armament of four 7·62 mm. machine guns, a range of 400–450 miles and a service ceiling of 30-32,000 feet. A later version had a 750 h.p. engine and a slightly larger wing span and length, but it only increased the top speed by some 5-10 m.p.h. The I 153 had an even more powerful engine, at first the 850 h.p. M 62 and later the 1,000 h.p. M 63. This gave it a top speed of 265-270 m.p.h. and with extra tanks, a range of over 500 miles. But the performance was still inadequate against contemporary British and German low wing monoplane fighters such as the Hurricane and the early versions of the Me 109. The I 16 designed by Nikolai Polikarpov in 1935 was more advanced. This low-wing monoplane single-seater fighter had, at first, a 750 h.p. engine of Gnome-Rhône inspiration or alternatively the M 25 built in the USSR under licence from the U.S. Wright engine firm. A later version, the I 16B was powered by the 1,000 h.p. M 62 engine and was extensively used in the early combats against the Luftwaffe both in Spain and the Germano-Soviet war; though it was faster than the I 15 (the latest version exceeded 300 m.p.h., but some had a top speed of about 280 m.p.h.) it was still greatly inferior to the contemporary Me 109 and the British Hurricane. The TB 3 four-engined bomber was too slow to be used in daylight operations and its engines needed more power. The SB 1 and SB 2 bombers, so-called fast bombers, were to need strong fighter escort and were inferior to the comparable German twin-engined bombers such as the Heinkel 111 and Dornier 17, both of which also needed strong fighter escort in daylight operations, except where there was little or no local fighter opposition. The fact is that the Soviet and German conception of a fast medium bomber which could attack its target effectively in daylight with little or no fighter escort was one of the many aeronautical pipe-dreams of the 1930s, a pipe dream shared by many British backers of the medium twin-engined Blenheim bomber.

All the same by the end of 1937, the output of the Soviet aircraft industry had reached, broadly speaking, that of the German, British and U.S. production plants. At that time, the USSR was making over 4,000 aircraft a year of which some 2,000 were single-engined fighters, nearly 1,000 were twin-engined bomber machines and some

200 four-engined bombers. But by then, Stalinism had cast a major shadow over future air development. This was the purge of the Soviet armed forces and the Communist political hierarchy. In December 1934, Kirov, the Communist Party leader was assassinated and already the period of the purges had begun. What did these mean to the Soviet Air Force? This and the performance of the Soviet Air Force in its first campaigns in Spain, Finland and against the Japanese Air Force on the China mainland are discussed in the next chapter.

CHAPTER 2

PRELUDE TO WORLD WAR

THE five years which preceded Operation Barbarossa, the code-name for the German attack on the USSR on July 22nd, 1941, were marked by three fairly large-scale bloodings of the Soviet Air Force in combat and also by large-scale internal purges in the ranks of its air staffs, air units, aircraft industry and air regiments. The purges had begun before the end of the Second Five Year Plan and there are conflicting views as to how the Soviet Air Force first came to be involved in them. The liquidation of military elements to strengthen the hands of the body politic has a long list of precedents throughout history. Stalin's repressive police measures had by the 1930s become a cardinal feature of Soviet life. It did not require the Khrushchev statements to the 21st Party Congress in Moscow in 1956 to divulge the Georgian's many tyrannies.

However, to do Stalin justice, there are grounds for believing that the Soviet Far Eastern Air Army was concerned in the period 1936-37 in a plot, no doubt well intentioned, to rid the Soviet Union of Stalin. The chief political administrator of these air units was Ian Gamarnik, bearded, patriarchal, ascetic-looking and ambitious. It would seem that he and the Commander-in-Chief of the Air Army, General Lapin, planned to use an air division of some 100 TB 3 four-engined bomber planes to drop anti-Stalinist leaflets urging Red Army troops to rid the country of Stalin and his confederates. The plot misfired, the conspirators were arrested in the summer of 1937 and Ian Gamarnik committed suicide in his office in the Kremlin. There is a view that this Far Eastern Air Force plot was one of the mainsprings that prompted the Soviet Army and Air Force purges of 1937-38. In these purges a very high proportion of the officer class in both the Soviet Army and Air Force suffered dismissal, imprisonment and execution. Right at the top, three of the five Soviet marshals who had been newly created in 1935 were purged, and they included the Red Army Chief of Staff, Marshal Tuchachevsky, known to be a supporter of the development of long-range transport aircraft for parachute operations and of long-range bomber units for strategic bombing of enemy industrial targets. Soviet Air Force leaders who were victims of the Stalinist purges included General Alksnis, the Commander-in-Chief, General Khripin, also a great believer in long-range bomber attack, General Lopatin, C.-in-C. of the Leningrad Air Command and General Muklevich, a senior tactical air force commander. Later, after the main purges of 1937-38, Air Generals Pavlov, Klimovksy, Korobkov, and Smushkevich, the latter also a Commander-in-Chief of the Air Force, were all liquidated. In addition at least half the air divisional and air brigade commanders, most of them colonels, were victims of the Stalinist purges in the years that preceded the Second World War.

Nor was it simply the officer ranks of the air force that suffered. At TSAGI Chaplygin, one of the senior aeronautical engineers lost his post to Kharlamov, technically a nonentity, but politically reliable from Stalin's point of view. And in the general ranks of the aircraft industry there were thousands of dismissals ranging from filing clerks to senior engineers. Among the top aircraft designers Kalinin, whose swept-wing aircraft designs were very advanced, was shot following a crash in a test flight of one of his planes carrying four party members. Tupolev, who was to become the chief post-war designer of Soviet heavy long-range aircraft, was sent to Siberia under the shadow of execution. But he survived to earn many a Stalin prize for his jet aircraft and many other marks of aeronautical honour.

It is not hard to evaluate the influence of Stalin's purges on Soviet air power. They affected not only a high proportion of staff officers of all kinds but also the rank and file of N.C.O.s and men. In both the German and Japanese General Staff appreciations of the purge effects it is stated that Soviet air leadership suffered as a result of the loss of so many senior officers, especially in the Finnish War and the first year of the German-Soviet War. One immediate effect of the Stalinist purges was the general atmosphere of mistrust, uncertainty and suspicion which they engendered in a country already dominated and harassed by the secret police and the Communist Party. Another was the extra care taken to choose the Soviet air contingents despatched to the Spanish Civil War in the summer of 1936. Ordinary

Soviet pilots were not considered trustworthy, despite their previous political vetting before they joined the Soviet Air Force. A particularly severe standard of political reliability was set. This in practice meant that a high proportion of the Soviet pilots sent to Spain were of senior rank, i.e. captain, major or colonel. The combat losses suffered in Spain in the period 1936-38 among senior experienced pilots added to the purge losses on the home front clearly weakened Soviet air leadership.

It was in the spring of 1936 that Stalin sent Berzin, chief of his military intelligence to Spain to study the problems of military intervention. At about the same time members of Goering's air staff were doing the same on behalf of the German Condor Legion. In the summer of 1936, beginning in July, the first units of the Soviet Air Force, numbering some 200 pilots and between 2,000 and 3,000 ground personnel left Odessa by sea in the ships *Rostock*, *Neva* and *Volga* en route for the Spanish ports of Alicante and Cartagena on the south-east coast of Spain The aircraft travelled to Spain in crates and so there was a delay in assembling them and getting them into action against the contingents of the German and Italian Air Forces. The Soviet fighter planes were initially mainly I 15 and I 153 biplanes, built at factories in Gorki, Vorenezh and near Moscow. But in 1937 the Soviet squadrons were to receive some faster monoplane I 16 fighters. The bomber planes were chiefly SB 2 twin-engined machines, so-called "fast bombers" (*skorostny bombadirovchik*). The reconnaissance planes were nearly all obsolescent R 5 biplanes and there were also one or two Sturmovik units equipped with rather obsolescent planes. The most striking omission was the absence of a force of Soviet transport aircraft which like the German Junkers 52 in the Condor Legion could have done much to fly key troops, spares, fuel and bombs as well as maintain wireless communications and improve the mobility and flexibility of both ground and air forces. But at the time of the Spanish Civil War, the Soviet air transport units were almost at the nadir of their pre-war fortunes. The TB 3 four-engined bomber transport was being phased out of production, Lisunov, designer of the main (twin-engined) Soviet transport machine of World War II, the LI 2, was still working on the blue-print of the machine and was to be sent on an extended course to the Douglas air plant in the U.S.A. Finally a number of Tupolev's prototype transport and passenger planes were crashing or otherwise failing, amongst them the well-known six-engined Maxim Gorky.

We have seen that the Soviet air contingents did not lack senior officers. Indeed the 1936 squadrons in Spain were commanded by General Smushkevich, deputy to the Soviet Air Force Commander-in-Chief, General Alksnis, whom he was to succeed when Alksnis was purged. The Soviet twin-engined bomber units were commanded by General Denisov who had a high reputation. Both Smushkevich and Denisov were to be Stalinist purge victims and so their valuable operational experience in Spain was lost. But in general the air operational record of the USSR in the Spanish Civil War was not impressive, due mainly to the superior training of Luftwaffe personnel and the superior technical performance of the German fighter aircraft, not so much the biplane Heinkel 51 as the new Me 109 which had a top speed of some 350 m.p.h. and could outclimb though not outmanœuvre Polikarpov's I 15 and I 153 fighters. However, the Soviet Air Force had some two years of useful operational experience during which nearly 1,500 Soviet aircraft were engaged, though in any one month not more than 400-500 of these were both crewed and serviceable. The bulk of the planes, i.e. about 1,000, were single-engined fighters of which some 500-600 were I 15 or I 153 biplane machines and the balance the more modern and faster low-monoplane I 16 fighter. Between 200 and 250 were SB 2 Tupolev medium bombers and the remainder nearly all R 5 biplane reconnaissance machines. Finally one regiment of some 40 Sturmovik RZ type aircraft took part. Also fighting for the Spanish Republican forces were air elements from France, Holland, Britain, the United States and Czechoslovakia. This international air force complicated the problems of servicing but gave the Soviet Air Force ground personnel varied experience. Some of the initial Soviet air operations were well executed and effective for instance against Majorca and Zaragossa in 1936 and 1937. This was the period when there was no German Me 109 fighter opposition. Nevertheless the wastage in Soviet planes, i.e. the crashes, the losses in combat and the low serviceability following intensive operations, soon reduced the military value of the Soviet air squadrons. But, as repair and assembly factories and bases were developed at Barcelona, Valencia, Cartagena, Alicante and Albacete, there was a revival of front-line operational strength. The one Soviet air operation which made the headlines was the attack by a force of some 25-30 SB 2 Tupolev bombers on the German cruiser *Deutschland* off Zaragossa on December 28th, 1937. The damage inflicted brought an excessive number of decorations to the Soviet pilots who took part. They were, it was stated officially, "for the excellent execution of special services to the Soviet government".

Encouraged by this and other local successes the Soviet authorities shipped far more fighters and bombers to Spain. At the beginning of 1938 a force of some 250 aircraft was intensively engaged on the Barcelona and Madrid fronts. Other forces were engaged in that year fighting over the battlefronts at Cordoba, Seville, Tervel, Toledo and Salamanca. The I 16 had some successes mainly against Italian opposition and the Natasha R 5 biplane machine, powered with a 500 h.p. and later a 750 h.p. engine, did useful armed reconnaissance operations in a campaign in which there was no radar used on either side to help fighter interception.

But in the summer of 1938 the arrival on the Spanish scene of some 50 Me 109E's in the German Condor Legion presented the Soviet Air Force units with dangerous opposition and in the autumn of that year Stalin wisely decided to withdraw the Soviet air squadrons. By the end of 1938 there were virtually no Soviet planes engaged in combat in Spain.

What lessons did Stalin and his air leaders absorb from the Spanish Civil War? The most important one was the failure of the Soviet aircraft industry to produce a modern fighter in the class of the German Me 109 or the British Hurricane and Spitfire. It is true that during the Spanish War, Soviet designers Lavochkin, Mikoyan, Gurevich and Yakovlev were designing and testing two or three new monoplane fighters which promised better things. The I 17 fighter exhibited at the Paris Air Exhibition of 1938, with an armament of four machine guns and one engine-mounted 20 mm. cannon and a top speed of over 300 m.p.h. promised to close the technical gap. But apparently the Soviet aircraft industry was then unable to cope with mass production of a machine of such complicated structure. The second lesson from Spain was the relative ineffectiveness of much of the Soviet bombing, an ineffectiveness which was not to be unique in modern bomber air forces used on such a relatively modest scale and without the help of pathfinder forces or some other target control system.

It would be interesting to know how far Stalin was influenced by the relative failure of his air force in Spain in taking the decision to make a temporary pact with Hitler in 1939. Like Neville Chamberlain at Munich, Joseph Stalin was playing for time and some of that time was desperately needed for the newly promoted Soviet aircraft designers, such as Ilyushin, Lavochkin, Mikoyan and Gurevich, to make up the technical leeway in aircraft design. The disastrous Soviet air operations against the Luftwaffe in the summer of 1941 were concrete testimony of how large this leeway was.

The Spanish Civil War achieved prominence as a testing ground for the latest German, Italian and Soviet planes. But the Soviet air operations over China and Manchuria in the 1930s which served the same function were rarely reported at the time. And yet the first blooding of the new Soviet air squadrons had been back in 1929 in minor air actions against Japanese- and Chinese-flown aircraft over Outer Mongolia. But then China in particular and Asia in general rarely obtained their due headlines in the Western press until after Pearl Harbor, indeed perhaps not until the Korean War broke out in 1950.

The chief Soviet air operations in Asia in the period prior to the Second World War were carried out by elements of the Far Eastern Air Army in 1937-39 on the fringe of the Sino-Japanese War. There was less overt interference than in the Spanish Civil War. Up to the end of 1937 the Soviet General Staff confined itself to training and equipping a number of Chinese air squadrons totalling some 150-200 aircraft. They comprised mainly I 15 and I 16 fighters and SB 2 twin-engined bombers. But in 1938 there were border clashes in Manchukuo between Japanese and Soviet ground and air forces; and so a "volunteer" force of four fighter and two medium bomber units, some 250-300 aircraft in all, was despatched to China, on the pattern of the Soviet intervention in Spain two years previously. This modest tactical air force was attached to the First and Second Red Armies of the Soviet Far Eastern land forces and up to the summer of 1939 was intermittently engaged in combat with Japanese army air units. When the Japanese ground forces attempted to cross the Mongolian frontier in May 1939 they were strenuously opposed in the air by the Soviet tactical air units. The air battles, near Buir Nuir, Nomonkhan and Khalkin-Gol, were intensive and according to Soviet sources the Japanese lost nearly 200 planes in the next two months; losses on the Soviet side were said to be about fifty. No doubt these figures are as wide of the mark as were subsequent air claims by Japanese, Soviet, British, U.S. and all other air forces. In all the 1939 campaigns against the Japanese Air Force, the Soviet airmen or their communist leaders claimed to have destroyed 600 enemy planes in combat. These claims, however false, indicate that some intensive air fighting must have taken place.

Two noteworthy types of air operation were carried out in this Far Eastern air campaign of 1939. The first was an attempt to use head-on ramming tactics by three Soviet fighter pilots Kustov, Skobarikhin and Mashnin who achieved much national newspaper fame in the USSR for these novel tactics. The second was the emergency use of three or four squadrons of TB 3 four-engined bombers to drop troops and supplies to Soviet ground forces cut off by the Japanese. It was only the shortage of long-range bomber-transport aircraft which prevented the Soviet Air Force from carrying out operations of this type much more often in the Second World War. The air campaigns between the Soviet Air Force and the Japanese Air Force in 1939 were not marked by air superiority on either side. Both sides concentrated on attacking ground targets such as troop concentrations, airfields and road and rail communications and neither side met continuously with strong local air opposition. The absence of radar and the primitive early warning systems in use must have made regular interception difficult for both air forces.

The Soviet attack on Finland in November 1939 might well have developed into a World War with Germany and the USSR ranged on the same side. The subsequent train of swift military happenings in 1940 and 1941, between the German invasion of Norway in the spring of 1940 and the Wehrmacht's attack on the USSR in June 1941, obscured this possibility. If Sweden had facilitated the passage of British and French "volunteers" through her territory to fight in the Finnish winter war of 1939, to the point where Finnish resistance could have prolonged the struggle to the spring of 1940, who knows if the Germano-Soviet pact of the summer of 1939 would not have been given real military teeth and so presented Britain and France with the combination of Germany and the USSR in the fateful summer of 1940?

But in the event the Soviet-Finnish War was to last only from November 1939 to February 1940. It produced disappointing results for the Soviet General Staff and encouragement to Hitler's pigeon-holed plan to pounce on the Soviet Union when the moment was opportune.

Amongst the Soviet air failures in the Finnish campaign, that of their much publicised parachutist arm was perhaps the most striking. The lack of proper equipment, which included even a shortage of parachutes, of skilled navigators and of experience in large-scale parachute operations resulted in largely abortive operations by about 5,000 Soviet parachutists. It is estimated that less than twenty per cent of this force dropped and fought in organised fashion, that is were dropped correctly in the allotted zones near the Mannerheim Line and Petsamo and were able to fight their way collectively towards their objectives. But the accuracy of Finnish sharpshooter riflemen, their clever snow camouflage, the failure of Soviet radio and ground communications after the parachutists had landed, made the hundreds of sorties by the TB 3 transport planes abortive in the main. Like subsequent air transport planes in other air forces, the Soviet aircraft were to drop supplies of ammunition intended for their own parachutists into the lap of Finnish ground troops who quickly learnt the Soviet visual signals and directed the Tupolev four-engined supply planes away from the isolated and uncoordinated groups of Soviet parachutists. In the light of their inexperience and lack of good navigators and navigational aids, so crucial in air warfare carried out in the depth of a Finnish winter, the Soviet failure in their first essay at large-scale combat parachutist operations is not surprising. It was certainly discouraging for men like General Gorbatov who had hoped for airborne troop successes against the relatively light opposition.

Certainly the air opposition to the Soviet bomber and fighter squadrons did not provide much of a test for the two air armies deployed in the Finnish campaign. The Finnish Air Force had no radar to give them early warning of Soviet raiders and their air force, though well trained and heroic, had a harlequin mixture of obsolescent European aircraft, such as the Soviet Air Force had flown in the Russian Civil War at the end of the First World War. While individual Finnish and foreign volunteer pilots, mainly British, fought with distinction and success there was little sustained Finnish air effort and so for the Soviet Air Force there was no problem of establishing local air superiority from the first days of air combat.

On the second day of the Soviet-Finnish War, formations of up to 50 medium bomber SB 2s escorted by the I 16 fighters attacked a wide range of objectives at a convenient bombing height and were virtually unopposed on most of the sorties. After the bombing of objectives at Turku, Lahti, Viipuri, Helsinki, Kemi and elsewhere at about 1,000 feet, the escorting Soviet fighters were able to come in low and ground-strafe at will. Inevitably Finnish civilians suffered as much as the military in the hundreds of bombing raids which took place in the next two or three months. President Roosevelt's appeal at the time for Soviet consideration of civilian casualties and damage was a curious failure to recognise the limitations of all bombing aids, Soviet or other, in use in 1939. And if Finnish schools, churches, hospitals and homes were reduced to rubble by the 250 lb. and 500 lb. bombs decanted by Soviet TB 3 and SB 2 machines, this was not only the inevitable bombing pattern of the time but very much the shape of things to come during the next few years in Britain, France, Germany and Japan.

The Soviet Air Force in Finland did not have to cope with the strong air opposition it had encountered in the Spanish Civil War and in the air fighting over Mongolian and Manchurian territory. But despite the lack of serious air opposition, the 7,000-8,000 tons of bombs dropped in this 1939-40 winter campaign, mostly by Soviet twin-engined medium bombers, did not achieve any major successes in support of the Red Army ground forces. The average-sized attack comprised some 30-40 SB 2 bombers or a similar-sized mixed force of SB 2 planes and the new longer-ranged faster twin-engined DB 3 bomber. There were some Soviet bombing successes, of course, for instance against Finnish shipping targets at and near Turku, against the main railway lines from Kemi and on military and air targets in the Helsinki area. These came mainly from a series of repeated attacks on the same target. In some cases the SB 2 formations returned on more than thirty occasions to the same objectives.

On the whole the two Soviet Air Armies engaged in the Finnish winter war must have been disappointed with what they achieved. Certainly Soviet incendiary bombs, so-called Molotov Breadbaskets, reached the headlines of the world's press when these hollow cylinders containing about one hundred small magnesium incendiary bombs were scattered widely. They became the chief cause of the damage and distress suffered by the Finnish civilian population, many of whom lived in wood-built homes particularly vulnerable to incendiary attack, much as the matchwood homes of the Japanese were in 1944-45, when they were subjected to the incendiary attacks of U.S.A.A.F. B 29s. But in the business of modern air operations, the Finnish War did not advance Soviet Air Force tactics, standards or reputation. The operations by over 2,000 planes engaged between November 1939 and the end of February 1940 revealed fundamental weaknesses which could not be remedied in time for the early crucial part of the Germano-Russian War. Wireless and radio-telephone contact in the tactical air sorties was poor and spasmodic, standards of blind-flying were inadequate, liaison between the Soviet Navy and the Air Force on the bombing of naval targets left much to be desired and, as in other air forces of the day, the use of the aerial torpedo had been little developed.

The Finnish climate and terrain were to blame for some of the Soviet air failures. The closely wooded country made it easy to camouflage many of the targets and the conduct of close support air operations in the winter conditions called for a sustained skill by individual crews which the Soviet Air Force did not possess. In these difficult conditions of snow, low cloud and ice-bound aerodromes a high crash and flying accident rate was inevitable, despite the light air opposition. The two Soviet Air Armies lost over a quarter of the aircraft engaged. But perhaps the casualties were well worth the losses sustained. Two years later, and again in 1942, Soviet air squadrons had to operate intensively in just such winter conditions in desperate defence of Moscow, Leningrad and Stalingrad. They showed themselves much better able to fly in these combat conditions than was the Luftwaffe, which was more used to blitzkrieg successes in Goering's summer weather.

If there was slow progress technically and tactically in the four or five years that preceded the Second World War, the expansion of the Soviet aircraft industry and of the Soviet air brigades and regiments was an achievement of which the Kremlin could be proud. Front-line operational strength, which had stood at about 1,000 aircraft in 1929-30, had increased to nearly 5,000 by 1938. Thus already the Soviet Air Force was as large as any other air force in the world. And by June 1941, when the Wehrmacht attack on the USSR was launched, its front-line strength had more than doubled, to a figure of well over 10,000 machines. Estimates by various Western intelligence departments during this period varied between 10 and 15,000 aircraft. This very impressive expansion in the operational strength of the Soviet Air Force on the eve of the Second World War was due to many factors of which clearly the most important was the expansion of the aircraft industry. This took place at plants and factories in European Russia, also beyond the Urals in or near Omsk, Tomsk, Khabarovsk, Kuznetsk, Chita, Magnitogorsk and Vladivostok. By 1938-39 the Soviet aircraft industry was roughly as great as that of Germany, Great Britain or the United States and much larger than that of Japan. It was turning out between 4,000 and 5,000 aircraft a year, of which about half were single-engined fighters, some 200 four-engined bomber-cum-transport planes, about 1,200 twin-engined bombers and the remainder mainly trainer aircraft. Single- and twin-engined transport and communications aircraft were being manufactured at a rate of 400-500 per annum. While the industry had suffered to some extent in the purges, a fresh group of designers could now be given more scope. They rose to the occasion and more than filled the gaps created by the blood-letting of Stalin's political policemen. There was for example, Semyon Lavochkin who had been an important designer in Kalinin's bureau. The latter's liquidation gave Lavochkin the chance to produce successful fighter designs of which the LA 5, 7 and 11 were the best known in World War II. He also worked with designers Gorbunov and Gudkov. Between them they engineered the LAGG series of single-seater fighters. And when Tupolev was sent into exile it provided extra opportunities for Petlyakov, whose PE 2 twin-engined plane was one of the outstanding successes of the Second World War. It was also in this period that Ilyushin was to design his IL 2 Sturmovik plane. This aircraft was test-flown in the 1937-38 period but did not reach the tactical air units of the Soviet Air Force until the end of 1939. Together with the PE 2 it formed the main equipment of Soviet tactical bomber regiments of the 1940-41 period. The PE 2 had a top speed of 330-340 m.p.h., that is it was almost as fast as the contemporary Hurricane British fighter or the United States P 40 single-seater fighter.

But the Soviet Air Force, despite the new LAGG fighters and Ilyushin and Petlyakov tactical bombers, was far from ready in 1939. The modernisation of the Soviet Air Force was a primary defence requirement of the USSR in 1940. The framework of the air force was large; it had the largest number of aircraft in operational units of any country in the world and an expanding aircraft production. But in radio equipment, mobility, air transport planes and in the technical performance of its fighter planes, it was sadly behind the German Air Force to which it was soon to be opposed. There had been an expansion in the Soviet Air Force air training organisation. New schools for both civil and military flying had been opened in the middle 1930s. Facilities were expanded at the flying training schools at Abakan, Barnaul, Zuchenko and Tomsk and new schools were set up for military flying at Ashkhabad and Bokhara in Central Asia. But the greatest expansion was naturally in the advanced flying training schools of which there were some 200 at the outbreak of the Second World War. The highest proportion of them were fighter trainer schools, concentrating on formation flying and aerobatics but often not reaching a high standard in either. There was little operational training done at these schools in long-range interception work as the pilots were intended almost entirely for tactical air and ground support operations. These tactical fighter trainer schools were located all over the Soviet Union, e.g. near Moscow, Odessa, Grozny in the Caucasus, Kuzhetsk on the Volga, Kustanar in Kazakstan and Kizil-Arvat in Central Asia. There were also special reconnaissance training schools at Kharkov, Orel, and Stalingrad and bomber training schools at Novosibirsk, Kharkov, Khaborovsk, Christopal and at least a dozen other centres. But at many of these advanced flying training schools there were major deficiencies in instruction, equipment and aircraft. The bomber crews were given little instruction in blind flying mainly because of the shortage of good radio equipment and radio operators. And so the heavy bombardment force formed in 1936 was almost bound to fail, especially in its operational tasks of night or bad-weather bombing. The Finnish War bombing failures were the fore-runners of similar lack of success in the early operations against German targets in the summer of 1941 when General Markoff who commanded a long-range bomber brigade of four-engined TB 3 planes reported that his air crews had been completely ineffective in their operations against Wehrmacht rail and road traffic advancing eastwards. Another Soviet air leader, General Kopetz, committed suicide in July 1941 when a month of combat failure had made him realise how ill-prepared and ill-trained his crews were for bombing operations.

This overall bombing weakness was accompanied by weaknesses in other fields. It was inevitable that much of Soviet tactical reconnaissance in the initial stages of the coming struggle with Germany would be ineffective, as at the reconnaissance air training schools there had been a lack of experienced army personnel to train the crews in accurate assessment of rail and troop movements and in quick recognition of Red Army ground signals. The standard of co-operation between Soviet tanks and aircraft was also bound to be lowered by the lack of good radio equipment in many of the tactical air squadrons.

Thus the Soviet Air Force was in many ways ill-prepared to meet the experienced, mobile, modern Luftwaffe to which it was to be opposed in Operation Barbarossa. It lacked a proper force of transport aircraft, operational experience of the kind the German Air Force had had, and its leadership was suspect. It had no effective long-range bomber force to use against the widespread vulnerable lines of communication of the Wehrmacht and it had no strategic fighter defence force worth speaking of. But it had a large aircraft industry, a large air force manned in the main by courageous and devoted air crews who were well used to uneven standards and inferior equipment and who were anxious to learn with that special brand of Slavonic exuberance and quick wit which was to mark the great progress of the Soviet Air Force in World War II. The early setbacks against the Luftwaffe in 1941 were inevitable, as were also the improvements in tactics, aircraft and equipment in the succeeding war years.

CHAPTER 3

BARBAROSSA TO BERLIN

APART from its numerical superiority and the courage and determination of its aircrews, the Soviet Air Force had no important operational factors in its favour on June 22nd, 1941, when Germany attacked the USSR in accordance with the Barbarossa plan. The Luftwaffe was superior in terms of combat experience, technical equipment, tactics, mobility and both air and ground crew training standards. The High Command of the German Air Force estimated the strength of Soviet air units in European Russia at about 7,500 aircraft out of a total front-line strength of perhaps between 10,000 and 12,000 machines. The Soviet air forces opposed to the Luftwaffe consisted of some 3,000 fighters, between 2,000 and 2,500 Sturmovik, medium and light bomber aircraft, 400-500 reconnaissance aircraft and some 1,500 transport-cum-bomber and liaison aircraft. The German Air Force deployed a strength of about 3,000 aircraft, including nearly 1,000 twin-engined bombers, some 350 dive-bombers, 600 single-engined fighters, 700 reconnaissance aircraft, including tactical and long-range planes, 60-70 twin-engined fighters and some 300 transport aircraft.

Perhaps the biggest and most important technical gap in the summer of 1941 between the two opposed air forces was in the crucial field of single-engined fighters. Nearly all the German Air Force squadrons were equipped with the Me 109 with a top speed varying from 360-380 m.p.h. according to the engine installed. A high proportion of Soviet fighter units was equipped with obsolescent I 15 and I 16 fighters with top speeds of 300 m.p.h. or even less. Even the new MIG 1, YAK 1 and LAGG 3 fighters coming into service in small numbers at the time had top speeds of only 340-350 m.p.h. and so were at a disadvantage in dealing with the faster German Me 109 F. It is not possible to make valid comparisons in all the main aircraft categories. The German Junkers 87 dive-bomber was an entirely different aircraft from the Soviet IL 2 Sturmovik and the latter was certainly as effective as the Germany machine when it had adequate fighter cover. The Soviet PE 2 and DB 3 twin-engined bombers were also not greatly inferior in performance and operational effectiveness to the German twin-engined Junkers 88 and Heinkel 111 bombers, though in this category many Soviet air units still had to fly obsolescent SB bombers which had not acquitted themselves particularly well in either the Spanish or the Finnish war.

In the case of reconnaissance aircraft and operations the German Air Force was clearly superior. Attached to nearly every German armoured division was a highly trained tactical reconnaissance squadron fitted with Henschal 126 or Focke-Wulf 189 machines, and these aircraft were flown by highly skilled crews, nearly all of whom had had experience in the Polish or French campaigns or even more recently in the Balkan campaigns against Jugoslavia and Greece. So indeed had the longer-ranged twin-engine reconnaissance squadrons flying Dornier, Heinkel or Junkers machines. Soviet tactical or longer-ranged reconnaissance suffered from obsolescent or inadequate equipment and inexperienced aircrews. The 1941 Soviet reporting of German troop and transportation movements, of ground signals and of covering camouflage was patchy and unreliable. These deficiencies were important, for the Soviet armed forces did not have the same efficient wireless reporting system that the Wehrmacht did and so relied more heavily on the results of aerial reconnaissance than did the German armed forces.

Neither of the opposing air forces had strategic bomber forces which could seriously disrupt or damage the basic war industries of their opponents. This was no great handicap to the Germans in the early months of the war but became a factor of increasing importance as the war dragged on. The Soviet Union had virtually the entire support of the growing Anglo-American strategic bomber forces in Europe; these not only did crucial damage to German war industry but eventually kept about half Germany's air resources away from the Eastern war fronts and pinned them down in permanent air defence of the Third Reich. The Soviet long-range bomber units in 1941 consisted mainly of the slow and obsolescent TB 3 four-engined plane which was to be used almost entirely for transport work or tactical operations in the battle area. Its successor, the PE 8, was not yet fully operational. When, in the spring of 1942, the Kremlin created a long-range flying command, its main bomber equipment consisted of twin-engine American B 25 Lend Lease aircraft, and of Ilyushin twin-engine DB 3F with a radius of action of between 500 and 600 miles according to bomb load. Thus both the German and Soviet Air Forces on the Eastern Front were to be almost entirely concerned with tactical ground support air operations throughout the war.

Other factors in favour of the Luftwaffe were the greater experience and flexibility of its unit and air corps commanders. Not only were Soviet Air Force commanders hampered occasionally by Communist Party interference with the course of operations, but many senior officers who had had experience in the Spanish and Finnish Wars and against the Japanese Air Force had been purged in the two or three years before the Barbarossa attack was launched. And the Luftwaffe airfield organisation and supply position were superior to that of the Soviet Air Force. The lack of motor transport and transport aircraft for moving Soviet air units was acute in the summer of 1941.

But the most important factor in the early weeks of the campaign was the fact that the Germans succeeded in their surprise mass attacks on Soviet Air Force bases in Poland, the Baltic States and the Ukraine. The absence of radar and the entirely indifferent dispersal facilities, aggravated the toll of Soviet aircraft destroyed on the ground. It was the inferior performance of Soviet fighter aircraft which was mainly responsible for the heavy losses in air combat against the Luftwaffe. In addition many useable aircraft had to be abandoned in the hasty retreats of the first few months. Within a matter of weeks, Vilna, Kaunas and Riga fell in the north, in the centre the Germans reached Smolensk and further south were attacking Kiev. Within two months they had reached the Dnieper and the Black Sea.

It is difficult to get reliable figures for Soviet aircraft losses during this period. The German Air Force High Command claimed to have destroyed over 3,000 Soviet aircraft in the first week of the campaign and a monthly average of nearly 3,000 aircraft up to December 1941. The Soviet Air Force certainly lost thousands of aircraft in the first few months of the campaign and reinforcements had to be rushed to the Smolensk, Leningrad and Kiev fronts by drawing on over 1,000 fighter and fighter-bomber planes in the Far Eastern Air Armies. Apart from this, aircraft from operational training schools were rushed into front-line service and transport aircraft from the Aeroflot civil air lines were turned over to the military. By the end of the summer of 1941, the Luftwaffe enjoyed fairly complete air superiority and could concentrate entirely on supporting the advancing German armies.

But even by the autumn of 1941 there were signs of Soviet air recovery and indications that the Luftwaffe had already bitten off more than it could chew. In the air battles above and near Leningrad, for instance, the First German Air Corps was too weak to carry out its task and had to be reinforced by elements of the crack Luftwaffe 8th Air Corps under General von Richthofen which was itself fully engaged and urgently needed in the stiff fighting near Smolensk. But Soviet air resistance in the Leningrad area was much more successful despite the Luftwaffe reinforcement. The German air attacks on Moscow first launched on July 21st had called initially for special reinforcement by Heinkel bomber units from France. But after two or three weeks the Luftwaffe stopped their attacks. Strong reinforcement of the Moscow area by Soviet fighter units and anti-aircraft batteries had so stiffened the air defences that German bombers were not having the desired effect. And in the Kiev ground battles German tank divisions were for the first time reporting strong and persistent Soviet air attacks by formations of IL 2 and PE 2 machines escorted by YAK and MIG fighters which were having an important effect on the ground battle. Finally by the autumn of 1941 many German fighter units were operating at half or two-thirds of their establishment strength and had to fight on throughout the autumn in this depleted state.

Within three to four months the operational strength of the German Air Force was reduced to some 2,000 planes for the whole of the Eastern Front, and this was at a time when there were a number of factors making for recovery in the Soviet Air Force and for giving them tactical air advantages. By making maximum use of transport aircraft and evacuation by rail, the elements of the Soviet aircraft industry in Western Russia were successfully removed from places such as Leningrad, Moscow, Kiev and Kharkov to production centres well outside the range of German bombers. It may be of historical interest that when Hitler issued his directive on Barbarossa on December 18th, 1940, it contained a clause that the Luftwaffe was to eliminate Soviet industry in the Urals if necessary. He seemed unaware of the range limitations of German twin-engined bombers.

This successful removal of industry to the East and the increasing despatch of aircraft from America and Britain were to give the Soviet Air Force's a superior aircraft supply position for the rest of the war. While the Luftwaffe was ill prepared to operate in the conditions of a Russian winter, the Soviet Air Force had had long experience in the art of keeping airfields serviceable under snow conditions, preventing engines from freezing up and fitting air and ground crews to face Arctic conditions. While the Soviet Air Force had warm clothing in abundance, many of the Luftwaffe personnel froze pending the results of Goebbels's appeal to the German public for furs and woollies. And as the Luftwaffe concentrated its strength for the autumn assault on Moscow, the Soviet Air Force was using a network of modernised and better equipped airfields near the capital, its fighter units on this front were more extensively equipped with the more modern YAK 1 and LAGG 3 fighters and they were operating close to supply air depots near Moscow so that the problem of replacements and spares was less acute than before.

In order to mount an air offensive on the Moscow front, using nearly 1,500 planes, the Luftwaffe had to seriously weaken the air corps operating on the Leningrad and Kharkov fronts. The German air squadrons were soon heavily depleted by the stiffening Soviet air opposition supported by a great concentration of anti-aircraft batteries. The conditions of poor visibility made much of the German tactical bombing and reconnaissance work ineffective. Crash rates were high. The replacement situation deteriorated even further and by December 1941, the German Air Force strength on the Moscow front had been reduced to about 500 planes of which usually less than half were serviceable. In contrast the Soviet Air Force was able to keep over 1,000 planes flying on the Moscow front, though when the Red Army counter-attacked early in December the Soviet Air Force could claim little credit for the recapture of German-held territory near Klin, Tula and Kalinin. As a German communiqué of the time put it, "methods of warfare in the East were conditioned by the arrival of the Russian winter". This meant that the weather precluded large-scale sustained tactical air operations on either side.

Early in 1942 an air development occurred which has been given little prominence in the military annals of the Second World War. In order to meet the growing needs of the Wehrmacht forces in the Mediterranean one of the three German air fleets operating on the

Eastern Front against the Red Army was withdrawn permanently. To this must be added the Luftwaffe need to build up its air defence forces in Western Europe, in Germany, Holland, Belgium, France and Denmark in order to meet the growing threat of the Anglo-American bomber offensive. From now on German air units on the Eastern Front were to have a lower re-supply priority and nearly all the freshly formed units were to consist of day or night fighters defending the Third Reich. Less than half the total air units of the Luftwaffe were now available for operations on the Eastern Front. Military historians generally take the Soviet victory at Stalingrad and the British victory at Alamein as the turning points of the war in Europe, but the turning point of the air war on the Eastern Front must perhaps be placed some nine months earlier.

By the spring of 1942 there were many signs that the Soviet Air Force was achieving air parity with the Luftwaffe. When Soviet ground forces drove the Germans out of Rostov in November 1941, Soviet Air Force tactical support was an important contributory factor. Two Ukrainian air armies totalling over 1,200 machines were used on this front. And at the turn of 1941-42 when the Red Army recaptured Mojaisk and Kaluga two further Soviet air armies of similar strength were engaged. But on both these air fronts German Air Force opposition was slight. And with the increasing numerical Soviet air strength came other factors such as greater experience in combat and growth of self-confidence and morale because of the recent victories and continuing modernisation of equipment, including the improvement of wireless communications. It was also in 1942 that over 2,000 fighter aircraft, mainly Hurricanes, Tomahawks and Airacobras reached the Soviet Air Force from U.S. and British sources. In the same year nearly 10,000 Soviet-built single-engined fighter planes reached Soviet front-line units and many of these machines were only slightly behind the German Me 109 in performance. These included an improved version of the LAGG 3 and the new Lavochkin 5 and Yakovlev 9 fighters on which the 12·7 mm. machine gun and the 20 mm. cannon were of improved design and performance. The IL 2 Sturmovik was now produced in a more effective two-seater version, with fairly heavy armament of two cannon and three machine guns, as well as rocket-firing equipment and improved armour and wireless equipment. This aircraft had become one of the most important and successful tactical air support weapons of the Second World War. The Pe 2 twin-engined plane was becoming more effective in both its bomber and reconnaissance rôles as the air initiative passed to the Soviet Air Force on the Eastern Front.

However, despite numerical superiority and all-round technical improvement the Soviet Air Force still had a long way to go before it achieved operational preponderance over the Luftwaffe, and from there final air superiority. This was illustrated in the intensive air operations which took place in the summer and early autumn of 1942 over the southern sectors of the Eastern Front from Kiev to the Black Sea. The German Air Force had concentrated a strength of some 1,500 planes and the Soviet Air Force mounted three air armies each of about the same size. But in the early May and June battles, particularly on the Kharkov front, the air operations by the Luftwaffe were more effective than those of the Soviet Air Force, which despite numerical superiority failed to achieve local ascendancy. In the May battles over the eastern part of the Crimea, despite their long prepared bases in the Kuban, their modernised equipment and their numerical superiority, the Soviet Air Force did little to prevent the eviction of the Red Army from the Crimea, any more than numerically superior British air forces in North Africa had been able to stay Rommel's advance in North Africa in the same period of the war.

Further north on the Voronezh front, however, the Soviet Air Force was more successful. The capture of Voronezh was a major target in the German offensive plan of the summer of 1942 which aimed also at the capture of Stalingrad, the securing of Caucasian oil and the vegetable and mineral wealth of the Ukraine. But strong tactical Soviet air forces, including a number of élite Sturmovik guards regiments not only blunted the assaults of German ground forces but compelled the Luftwaffe to operate at an intensity beyond its resources. The Soviet Sixteenth Air Army under General Rudienko distinguished itself particularly on this front in the summer of 1942.

One may well ask why the Soviet forces were unable to stem the German advance to Stalingrad and the Caucasus in the summer and early autumn of 1942. It was due mainly to a strategic error in the Soviet General Staff anticipation of the general direction of the Wehrmacht attack. The major Soviet ground and air resources were displaced northwards from Voronezh to Leningrad with a concentration in the Moscow area. The pick of Soviet air squadrons was kept to support the armies on the central front because the Soviet General Staff had anticipated that the direction of the new offensive would be towards Moscow; and so when Goering concentrated 1,000 planes to support the Wehrmacht attack—about half his total air resources on the Eastern Front—Soviet Air Force reaction to this concentration was ineffective until well into October. In September, when Hitler publicly promised the Germans that he would capture Stalingrad it seemed that such a success was likely, and by October 14th, the Germans had pressed the Soviet defenders of Stalingrad so close to the Volga that it seemed they could no longer hold the city. But by November there were several new air factors operating, all favourable to the Soviet cause. Soviet bombers and transport planes of the recreated ADD Long Range Air Command under General Golovanov had ferried down tens of thousands of ex-parachutists and guards, rifle and artillery personnel drawn from élite units in the Moscow area. The rapidly weakening Luftwaffe was completely incapable of interfering with this air ferrying of fresh troops and supplies. This was because powerful forces of IL 4 and IL 2 and PE 2 bombers, strongly escorted by some of the new Yakovlev and Lavochkin fighters had been concentrated under the supreme command of Generals Zhukov, Vasilevsky and Voronov and had made devastating attacks both on the Luftwaffe's forward airfields within fighter range of the Stalingrad front and on ground targets in the area of the crucial juncture between the German Sixth Army and German 4th Panzer Army. While the Soviet Air Force was able to throw in two new air corps totalling some 1,000 planes in October and November, the German Air Force efforts to support von Paulus and the German Sixth Army were inevitably poor. The weakly escorted Junkers 52 air transport units supplying the German Sixth Army had suffered such heavy losses, particularly at the hands of the new Lavochkin 5 air regiments, that they had to be withdrawn from these operations. Luftwaffe fighters were too urgently needed in the Mediterranean and the air defence of Germany in the autumn of 1942 to be switched to the Stalingrad front. But even before this phase, and despite the use of Heinkel 111 bombers to boost the German air lift, instead of the required 300 tons a day to supply von Paulus, the Luftwaffe achieved a figure of just over 100 tons a day. And in December 1942 and January 1943 the German Air Force casualties rose when ice, fog and drifting snow on the front made intensive air supply operations extremely hazardous and the efficient servicing of machines virtually impossible. In December 1942 and January 1943, half the German aircraft losses were due to the weather and in that period many of the German air transport units were operating at between ten per cent and twenty-five per cent of their normal strength.

When von Paulus's Army surrendered at Stalingrad in February 1943, it seemed as though the influence of German air power on land operations on the Eastern Front was virtually ended. The Luftwaffe certainly failed to provide more than token air support to meet most of the new Soviet ground offensives. In the north on the Leningrad and Moscow fronts while two Soviet Air armies, the 13th and 15th, were each increased to a strength of nearly 1,000 aircraft, the Luftwaffe air corps, depleted since October 1942, had no more than 200-300 machines fit for operations. This great disparity was typical of the war with one or two exceptions. The Anglo-American strategic bomber plan devised at Casablanca at the end of 1942 had aggravated the German air situation, for it meant that throughout 1943 and up to the summer of 1944 some three-quarters of all Germany's fighter aircraft resources were to be pinned down in the West as part of the permanent strategic air defence of the Third Reich. The Japanese Air Force now had its hands more than full in Asia and the Pacific, and Stalin could count on the support of continuing American long-range air assault on Germany and her allies. It is doubtful if any other air force in military history has been so free to indulge in the luxury of an undivided air directive to support ground forces, without looking over its shoulder at other air commitments, as was the Soviet Air Force in the spring of 1943. A single division of some 200-250 single-engined fighters was all that was retained for token local air defence in western Russia and the supply of fighter planes from the U.S.A. was more than enough to offset the effect of this modest commitment.

The disparity in size between the two opposing air forces in the last two years of the Second World War was large enough. While the Soviet Air Force never had less than 10,000 planes in its operational units and at peak moments such as the summer of 1944 and of 1945 had over 15,000 aircraft on front-line duties, the German Air Force on the Eastern Front rarely exceeded 2,000 aircraft. But Soviet numerical superiority was in a sense not the main factor. Large numbers of the best German pilots and aircrews were withdrawn from the Eastern Front to man German fighter defences in the West and the crews that replaced them were much less competent, not so well trained and lacking the morale boost provided by the operational successes which had attended German air operations in previous years. In the Stalingrad air campaign, the Luftwaffe had lost over 1,000 experienced air crew members who were virtually irreplaceable. Moreover the Luftwaffe had to make more and more use of obsolescent aircraft for close support work such as Henschal, Gotha and Heinkel biplanes withdrawn from training schools, at a time when Soviet aircraft were technically improved to the point where their performance was nearly equal to or equal to the most modern aircraft used by the Luftwaffe on the Eastern Front. Stalingrad had shown that the Soviet Air Force had large numbers of LAVOCHKIN 5 and YAKOVLEV 9 fighters fitted with engines of over 1,500 h.p. and capable of top speeds of over 350 m.p.h. In 1943 the LAVOCHKIN 7 with self-sealing tanks and a top speed of over 370 m.p.h. was also to be in service. Between 1943 and 1945 Soviet aero-engine designers had boosted power to over 2,000 h.p., thereby achieving in two years what had taken United States' engineers over four years. Such comparisons are inevitably unfair. But some of the rapid Soviet fighter developments which staggered the Western world in the early 1950s had clearly begun in the middle 1940s. The progress was there to be seen by the West, for at this time French pilots on the Eastern Front were flying Soviet Yakovlev fighters with top speeds of 375-380 m.p.h. and knew of a Lavochkin successor to the LA 5 and LA 7 which had a top speed of 400 m.p.h.

In the close support, ground-attack and medium bomber aircraft categories the Soviet Air Force could claim that by 1944 its native-designed aircraft compared quite favourably with their counterpart machines in either the Luftwaffe, the R.A.F. or the U.S. Army Air Forces. The Sturmovik IL 2 for instance was being phased out in some units and replaced by the improved version of it, the IL 10, designed by Ilyushin and fitted with a 2,000 h.p. AM 42 in-line engine, giving it a top speed of 335 m.p.h. and a bomb load of about a ton. It had both forward and rearward firing cannon and was capable of firing rockets. Incidentally the Soviet Air Force can rightly claim to have pioneered the use of rockets in air attacks on enemy armour. The Tupolov 2 twin-engined bomber had Shvetzov radial engines of 1,800 h.p. which gave it a top speed of over 350 m.p.h., a range of over 800 miles and a bomb load of about two tons. This modern tactical bomber also began to equip Soviet air regiments in 1944. And these technical aircraft developments went hand in hand with improvements in the Soviet Air Force supply and maintenance systems. In both spheres the Russian capacity for hard work and hard drinking and the ability to harness large-scale religious fervour to technical and scientific progress enabled the Soviet Air Force and aircraft industry to achieve far-reaching achievements much quicker than in other countries. These factors aided by the boosts of the highest Kremlin priorities explain much of Soviet progress in the present jet, rocket and atomic weapons age.

In the last two years of the Second World War a more flexible Soviet ground organisation was set up to meet the needs of the eighteen operational air armies each of which had its own local supply of aircraft spares, bombs, fuel, ammunition, airfield construction equipment and food. There were now more Soviet transport aircraft available to move these things from air depots in the rear and to move forward ground equipment and staff and repair depots with the advancing Soviet Army. By the spring of 1943 the air pendulum in all its aspects had swung violently in favour of the Soviet Air Force. Had it not been for an unexpected early thaw at that time which restricted Soviet supplies and ground communications, German air disasters and Soviet air superiority in this period might have been even more marked. As an indication of Soviet air potential at this time the Red Army in its recapture of Kharkov and Bielgorod had the support of no less than four air armies under General Novikov, later Commander-in-Chief of the Soviet Air Force, with a combined strength of over 6,000 machines, at the same time as two other armies, the 13th and 15th Air Armies, with a combined strength of 2,000 aircraft, were supporting the Red Army's assault on Rzev and Yyasma.

The opposing Luftwaffe units had usually less than one-sixth of the total of Soviet planes facing them and their aircraft were often manned by tired and dispirited crews and led by squadron commanders discouraged by defeat, low serviceability of aircraft, difficult airfield conditions and shortage of supplies of all kinds. In general, even before the summer of 1943, the Soviet Air Force had air superority over the Luftwaffe, for on many sectors of the front the German air squadrons could provide only a token tactical reconnaissance force, a few obsolescent biplane units for tactical air-support with a small number of inexperienced fighter squadrons to escort and patrol. This was but the illusion and semblance of air support. And yet here and there the Soviet Army and Air Force still failed to maintain the local initiative. Falling back on main supply depots at Nikolaev and Poltawa, the 8th German Air Corps was able to help von Manstein's ground forces to reoccupy positions on the Donetz river. But in the main along this and other fronts, Soviet Yakolev, Lavochkin and MIG fighters maintained constant patrols over Red Army bridgeheads and had little difficulty in neutralising the weakening German air power. Weakly opposed Soviet IL 2 Sturmovik planes and PE 2 light bombers caused havoc, dislocation and damage among German armoured forces and infantry in the Ukraine battles of the spring of 1943.

And southwards on the Kuban and Crimean fronts Soviet air successes were most marked in April 1943, though in the previous months the Luftwaffe had strained its resources to strengthen the 4th German Air Fleet on this front and transfer nearly 500 aircraft to Crimean airfields. But Soviet medium-range reconnaissance aircraft, mainly PE 2 and IL 4, had spotted the German air reinforcements, and General Vershinin, later to be Commander-in-Chief of the Soviet Air Force, was able to deploy over 1,000 planes in a successful series of attacks on German occupied airfields in the Crimea and on German shipping operating across the Kerch Straits between the Crimea and the Kuban. Again straining its waning resources the Luftwaffe put up an air effort of 400 sorties a day while the persistent and effective attacks of the Soviet Air Force involved operations of about five times the size, operations well within the scope of their local resources. In the air battles over the Kuban bridgehead the German Air Force failed to control the sea communications to the Crimea just as a month later it was to fail to control the sea communications between Sicily and the Tunisian bridgehead. And in each case it was to weaken its air potential for crucial tactical air battles in the coming summer campaigns. In these spring 1943 campaigns Soviet air units under Generals Krukov, Tupikov and Denisov distinguished themselves in a series of operations which revealed the all-round improvement of Soviet air regiments. Reconnaissance of the Crimea and Black Sea areas had been efficient and comprehensive and the follow-up torpedo-bomber attacks on German-Roumanian convoys had been well planned and pressed home with determination. Fighters had effectively protected the Soviet-held Black Sea ports and patrolled the areas across the Kerch Straits and the Kuban bridgehead as well as escorting Petlyakov and Ilyushin bombers effectively in attacks on German-occupied airfields in the Crimea.

Despite the handicaps of lack of reserves and ill-trained crews, the Luftwaffe maintained the offensive spirit and gathered its failing resources for the coming air offensive on the Central Front. By equipping Hungarian and Roumanian squadrons with the latest Junkers 88 bombers and the 109 fighters, by taking more obsolescent biplanes from training schools for ground attack units, by strengthening the German 4th and 8th Air Corps as far as limited resources permitted, the Luftwaffe was able to scrape together a force of just over 2,000 machines. Against this the Soviet Air Force had three air armies, the 1st, 4th and 16th, with a total strength of over 6,000 machines, and other air armies in reserve which could be thrown into battle if required. One of the major mysteries of the summer campaign of 1943 on the Eastern Front was the series of long-range night air attacks which Luftwaffe He 111 units opened up against Soviet industrial targets well behind the front. At this stage in the battle, it is hard to see how night attacks on the Soviet tank works at Gorki, the rubber plant at Jaroslav, the oil refineries at Saratov and the fuel depots at Astrakhan could have had much military effect, carried out as they were by forces of less than 50 twin-engined bombers, without the help of a pathfinder force or of radar navigational aids. Against these Heinkel night bomber attacks the Soviet Air Force used one or two recently created PE 3 night-fighter squadrons; but these were as ineffective then as they were a year later in defending the U.S. Flying Fortress temporary air base at Poltawa when it was attacked at night by Heinkel bombers. The absence of early warning radar and of ground controlled radar interception equipment was bound to make most Soviet night-fighter operations fruitless.

The last big clash of Soviet and German tactical air strength on the Eastern Front began on July 5th when the German Army on the Central Front went over to the offensive. By this time there were four Soviet air armies of nearly 10,000 machines ranged against the combined German, Roumanian and Hungarian air units of about

2,000 aircraft. The German High Command did little to mask the timing and direction of the new offensive and the intelligence resources of Britain did useful work in helping the Soviet to anticipate the attack at least two months before it began and so give them time to build up stronger land and air forces to meet it. In a crucial ten-day period from July 5th to July 15th the German Air Force made its last big tactical air effort on the Kursk-Orel front averaging roughly 3,000 sorties a day. But the Soviet Air Force reply was not only more than twice as great, it was also well directed. Using as many as 200 aircraft per operation, including strong forces of SB 3, DB 3, and PE 2 bomber planes, the Soviet Air Force carried out persistent attacks on Orel and its adjacent airfields and on rail and road targets in and near Briansk, Roslav and Gomel. That the German 9th Army penetrated only 9 or 10 miles in the first two days of operation "Citadel", as this offensive was called, was partly due to an intensive series of attacks on their ground forces by heavily escorted IL 2 Sturmovik formations. With the news of the successful Anglo-American landing in Sicily even Hitler decided to call off the attack. On July 15th the Red Army launched a counter-attack on Orel and at the same time a further assault on the lower Donetz front. In these operations documentary sources indicate that the Soviet Air Force maintained an average daily effort of 5,000 sorties, reaching the figure of over 10,000 on some days. The German Air Force in reply could achieve about one-quarter of this scale of effort. When Stalin announced on July 23rd that the German summer attack had collapsed he might also have announced had he wished that air superiority over the Luftwaffe was now virtually assured all along the Eastern Front.

One aspect of Soviet air domination in 1943-44 was the extent to which they could afford to divert bomber and transport resources to assist Tito's partisan forces in Jugoslavia. Thousands of PS 84 and IL 4 sorties were to be flown in operations of this kind. No doubt the Soviet High Command realised that the concentration of the Luftwaffe resources on the Kursk-Orel front in the summer of 1943 was to be the last of its kind. From now on the German Air Force had to split its weak resources in a fruitless effort to provide some form of regional air cover. Thus the Fourth German Air Fleet on the Southern Front was divided to cover the Stalino-Taganrog front and to stiffen the Soviet advance further north on the Kharkov-Poltawa front. Both Kharkov and Taganrog fell to the Red Army in August. In the next month Briansk and Smolensk were lost by the Wehrmacht, in October the Soviet capture of Nevel opened the way to the Baltic and in November Kiev fell to the Red Army. And now in the autumn of 1943 the Soviet Air Force was deploying twelve air armies, with a strength of some 350 regiments, that is, over 10,000 aircraft, against the Luftwaffe and because of the superior serviceability of Soviet Air Force units they frequently outnumbered German Air Force opposition by more than ten to one. For the next year or so, the log books of Soviet air operations were to record time and again attacks on German troops, airfields and road and rail communications with no opposition from Luftwaffe fighters and little anti-aircraft gunfire. The records of the Soviet 4th Air Army on the Smolensk and later Polish fronts were typical. Further testimony of the overwhelming air superiority of the Soviet Air Force from the autumn of 1943 to the summer of 1945 can be found in the published testimony of the French Normandie-Niemen squadrons flying YAK fighters under the command of Colonel (later General) Pouyade. In such conditions Soviet pilots could press home their attacks, maintain good flying formation and ensure effective air assault on the selected targets. Here was the ideal situation for perfecting close and speedy co-operation with Soviet armoured forces. Soviet claims that tactical air support arrived within half an hour of the request for it are borne out by German sources. There were of course lapses, the occasional failure to identify ground troops with consequent attacks by Soviet Air Forces on Soviet ground forces. Which air force was free of such occasional errors in the Second World War? And French pilots confirmed R.A.F. experience on the Murmansk front that the offensive spirit of Soviet fighter pilots made them liable to abandon close escort of bombers and go in search of air prey on their own private free-lance patrols.

The 1944 air story on the Eastern Front was a catalogue of Soviet successes in the air and on the ground. As the Luftwaffe committed more and more of its fighter resources to the air defence of Germany, so Soviet air operations encountered less and less opposition. They were able to support the ground advances all along the front from Leningrad to the Black Sea with the same freedom of action as that enjoyed by Anglo-American air forces operating in Western Europe in the summer of 1944. By the spring of 1944 the Red Army had advanced into southern Poland and the recapture of the Ukraine opened the war to Roumania. In this latter campaign General Vershinin's air armies included many units of the new YAK 3 fighter with a top speed of over 400 m.p.h. He also began to make more intensive use of specialised tactical reconnaissance squadrons whose efficiency was soon to enhance the effectiveness of the PE 2 and IL 2 tactical support bombers operating on the front. The Soviet summer offensive of 1944 thrust the Red Army deeper into Poland and the Baltic States. Vilna, Lublin, Polotsk and Omsk fell. Soviet fighter and ground attack units followed up the advance on airfields often ten to thirty miles from the front and this with no early warning system! They could now dispense with any form of local air defence and concentrate all resources on the close support of the Red Army advance. How well Stalin could have spared air support for the Polish underground uprising in Warsaw in August. How fantastic that 187 British and U.S. transport planes should have to fly air supply operations to Poles in Warsaw from Italian bases on the same nights that Soviet transport aircraft flew air supply to Tito's forces in Jugoslavia from south Polish bases.

And in October the Red Army captured Warsaw after taking Ploesti and its oil in September. It was in this month that Soviet bombers began a series of long-term and for them fairly long-range attacks on Budapest. It was towards the end of 1944 that the Soviet Air Force built up to an operational strength of over 400 air regiments with a combined strength of nearly 15,000 aircraft. And in the final air campaigns in support of the capture of East Prussia, Roumania, Bulgaria, Hungary and parts of Austria and Czechoslovakia it was a Soviet air Goliath against a weak puny myopic Luftwaffe David, myopic because the poverty of German air reconnaissance and ground communications was now making it almost impossible to see the way to devise concerted air or ground operations. In fact the Soviet Air Force did not play a crucial part in the final battles of the spring of 1945. It was the R.A.F. and the U.S.A.A.F. who hammered home most of the final air attacks on Berlin while the Red Army hastened across the Oder to ensure that the Soviet would occupy the German capital. Meantime the Soviet Air Force operated in support of the Red Army final advances on other fronts, operating fairly cautiously as though unaware of how weak German air opposition was in the last few months of the war.

There was naturally great Soviet intoxication after the final victory. Air General Rudienko said that the Soviet Air Force destroyed 75,000 of the 80,000 planes produced by Germany from June 1941 to May 1945. German records show that in a war in which air claims by all air forces had an indifferent factual basis, this particular Soviet claim deserves the aeronautical blue ribbon in the general military category of casualty and damage claims.

Looking back soberly at the operational record of the Soviet Air Force in World War II one finds that it is difficult to put into perspective. When the German Air Force was technically and tactically superior in 1941, the Soviet Air Force showing was generally poor. By 1943 the technical gap between the two opposing air forces had narrowed, but the numerical disparity was so great that the Soviet problem of gaining local air superiority was not a difficult one. The bulk of Soviet air operations were so lightly opposed from 1943 to 1945 that they rarely underwent the full test of stern air battle. Moreover a very high proportion of Soviet air operations were tactical short-range missions which do not call for the same complicated flying skills and staff work that long-range bomber or strategic air defence requires. At the end of the 1939-45 War one knew virtually nothing about Soviet air potential in these two important contemporary military fields. Although Soviet tactical air weapons were not far behind those in the German, U.S. and British Air Forces, in the use of radar, jet aircraft, long-range transport planes and guided weapons it seemed that the Soviet Air Force was several years behind in 1945. One of the great political and military questions of that period was, "How long would the Soviet Union take to close the technical gaps?" There can be few military prophets who predicted at the time that in the decade that followed the Second World War, the USSR would make the astounding progress it did in developing such a large range of modern aircraft, missiles and equipment in the new jet-atomic age.

CHAPTER 4

THE GERMAN LEGACIES

NO country in the world can claim that it does not learn and benefit from the aeronautical genius of other countries. Britain has flown and is flying U.S. designed planes and the United States is using British machines. British engines are installed in planes designed by Frenchmen, Dutchmen and other nationalities. Examples of international interdependence in aircraft and equipment could be multiplied to tedious lengths. Nevertheless the Soviet air debt to other countries was especially heavy between the two World Wars; the USSR drew particularly on German aeronautical genius at this time, and again in the post-war period of jet and rocket aircraft, guided missiles and long-range rockets.

Because of the poverty of Soviet resources in the early 1920s, Lenin had been compelled to look to other countries for air engineers, designers, technicians, flying instructors, money for creating passenger air lines, parachutes, bomber, fighter and reconnaissance aircraft, in fact for most of the air requirements of the newly created Soviet air squadrons. Italy, Holland, France, England and the United States all provided aircraft or aero-engines or equipment. But it was to defeated Germany that the new Soviet government turned to get most of its help in laying the foundations of the new Soviet Air Force.

The Treaty of Versailles, of which the air clauses were signed in June 1919, stipulated that Germany was no longer to maintain military or naval air forces. Her air personnel were demobilised and more than 15,000 aircraft and over 25,000 aero-engines were handed over to the victorious allies. In 1922 limitations were imposed on the size of the aircraft that Germany could construct. In such circumstances German air technicians and pilots who wanted a future in their chosen calling naturally turned to other countries. And so designers such as Claude Dornier, Ernst Heinkel and Hugo Junkers built or subsidised aircraft factories in Italy, Switzerland, Denmark,

Sweden, Turkey, but above all in the USSR. Liaison between Germany and the USSR was strengthened by a secret agreement reached in 1921 between the Red Army and the Reichswehr which provided Germany with aviation outlets in the USSR and Soviet staff and other training facilities in post-war pre-Hitler Germany.

And so in 1922 the Junkers firm built aircraft manufacturing facilities at the FILI plant near Moscow, where later on Tupolev's bombers were to be assembled, showing in their wing designs a marked Junkers influence. Heinkel built near Kharkov and Taganrog biplane fighters which were to equip some of the first air regiments of the Soviet Air Force. The Soviet air training schools at Kharkov and Leningrad were staffed generously by German flying instructors. Gliders and glider instructors of the German Luftsportverband found their way to the USSR and helped to put gliding in the Soviet Osoaviakhim flying clubs on a firmer basis. Germany also made a generous contribution of money, planes, personnel and equipment to the first Soviet civil airline Dereluft. Indeed the initials stand for German-Russian Passenger Airline and when the Company was founded in 1921 the original capital totalling some 800,000 marks derived almost entirely from German sources.

The German help to Soviet aviation continued well into the 1930s. Dereluft, which had in the early 1920s flown freight and passengers between Moscow and Berlin, extended its network to Leningrad in 1928. The Bayrische Motoren Werke (BMW) provided the prototype engines for some of the first Soviet-designed bomber and fighter aircraft. The bulk of German pilots and aircraft engineers who had moved eastwards to the Soviet Union in the early 1920s stayed on until the early 1930s when the advent of Hitler to power and the emergence of Goering as Commander-in-Chief of the new Luftwaffe naturally provoked an exodus and a return to the Third Reich fatherland.

Apart from the foregoing technical air aid, about a hundred Soviet officers were sent each year to Berlin for advanced military staff studies; Germany also sent some of her senior staff officers to the USSR, including General Werner von Blomberg, who was to become an influential figure in the development of the new German Wehrmacht in the middle and late 1930s. Amongst the senior Soviet officers who went to study in the German capital were Mikhail Tukhachevsky, later to become the Red Army Chief of Staff, and to be purged by Stalin, and George Zhukov, one of the Red Army's most brilliant generals in the Second World War and subsequently Minister of Defence until removed from that post by Nikita Khrushchev in 1957.

It is, however, difficult to estimate effectively German influence on Soviet military doctrine in the first formative decade or so of the Red Army and Air Force. The joint staff discussions and studies of the 1920s were occasioned by the Versailles Treaty restrictions and the 1922 Treaty of Rapallo between Germany and the USSR which had secret military agreements. These made for the joint military discussions and staff training. But there were inevitable parallels between Soviet and German military and air doctrine during that period resulting from sheer geography, which is so often the root of strategy. Both the USSR and Germany were great continental powers which would inevitably think of the air arm mainly as a weapon of tactical support for ground troops, as both countries would think militarily in terms of large armies fighting land-locked battles over a wide area. Certainly the initial organisation of Soviet air units into regiments and brigades and the relative unimportance of the Soviet naval air force may well have resulted in part from the representation of von Blomberg and his colleagues in Berlin. But in the final analysis the Communist Party and Stalin, and the technical limitations of the aircraft and equipment available, were much more important factors in shaping Soviet air policy than any joint German-Soviet staff discussions in Berlin and Moscow. In any event these discussions came to an end in the early 1930s, when both Soviet and German air doctrines like those of other nations, were in the tentative melting-pot stage. The real lessons of air power can only be hammered out in combat. It is likely that German air influence, in terms of strategy, tactics and doctrine was much greater when the Soviet Air Force was opposed to the Luftwaffe in battle, than in the hothouse of joint staff college debates and lectures, or of joint staff discussions.

And so it was in the Spanish Civil War and the four years of air combat from the summer of 1941 to the summer of 1945 that Soviet air thinking was most directly affected by the Luftwaffe. The Spanish Civil War impressed on the Soviet Air Staff the ineffectiveness and vulnerability of the so-called fast medium bombers (the SB 1 and 2 series) and the technical inferiority of the I 15 and I 16 fighters compared with the new Me 109 fighter. Their unhappy combat experiences in the Spanish Civil War taught the Soviet General Staff that the potential of the fast unescorted light or medium bomber, such as the German Do 17 and the British Blenheim was another of the uncertain air tenets of the 1930s. World War II experience was to show that only the unescorted fighter carrying bombs was to be generally effective in this rôle. Another of the major lessons which the Soviet General Staff must have absorbed was the importance of a large force of transport aircraft to support and achieve mobility and flexibility for the tactical air units. The Luftwaffe sent a useful force of Ju 52 three-engined transport planes, which not only ferried important contingents of troops into battle but also helped to maintain a high level of serviceability and flexibility in the German air units of the Condor Legion. But it was between June and August 1941 that the Soviet must have learned most about their own lack of air mobility and of the level of mobility with which a few hundred transport aircraft can endow a large air force. The manner in which the five German tactical air corps deployed in support of Operation Barbarossa switched their air squadrons from one sector of the front to another and kept right up with the rapid advance of the German Panzer divisions over distances of hundreds of miles, must have been both the envy and the admiration of the Soviet army and air staffs. German equipment, fuel, bombs and spares were moved to advanced bases almost as soon as Soviet personnel had left them. In contrast the retreating Soviet Air Force was caught flat-footed, and had to abandon hundreds of usable planes and dumps of bombs and equipment because it lacked adequate air transport resources. This lesson of mobility was learnt slowly by the Soviet air authorities. Even at the end of World War II, there was a shortage of air transport units to support the rapid Soviet advances of 1943-45. It was an improvised air transport lift, consisting mainly of Soviet bomber squadrons, which enabled the Red Army to rush reinforcing divisions from the Moscow area to the Stalingrad front in October and November 1942. Equally so it was an inadequate German air lift consisting of Heinkel bombers and Junkers transports in limited supply which failed to reach the minimum target of air supplies for von Paulus's surrounded Sixth German Army.

There is no need to stress that the Soviet military authorities have by now absorbed fully the lesson of the importance of air mobility in modern war. Modern developments in Soviet helicopters and jet transport aircraft are dealt with in detail in Chapter 11. To-day the Soviet armed forces air lift is probably greater than that of any other nation, particularly when one remembers that the resources of Aeroflot, the Soviet civil air lines, are completely at the disposal of the military formations. Moreover local self-sufficiency in aircraft fuel, bombs and equipment has been established as part of the Soviet counter to nuclear and thermo-nuclear attack which would probably be directed against its main cities and centres of rail communication. This reduces the overall problem of air supply.

If the importance of mobility was the main lesson that the Soviet General Staff imbibed from its operational experiences against the German Air Force, there were many other lessons to be drawn. Luftwaffe superiority in air combat tactics, in blind flying, in radar and in radio communications underlined the technical weaknesses of the Soviet air formations in these fields during the Second World War. At the end of that war the Soviet Air Force reviewed and revised many of its concepts of air power, both tactical and strategic. In order to do this and to catch up with the West in air technology, major technical advances were needed and needed quickly. Sound theory, it is said, is sound practice conscious of itself. To get sound practice in the new strategic air concepts of the jet, atomic, rocket, electronic age of the 1940s and 1950s needed major windfalls and legacies, in other words the large-scale use of Western technological advances. In defeated Germany the Soviet authorities found perhaps one of the largest and certainly the most advanced pools of scientific and engineering skills which any victorious nation has ever acquired in modern war.

It has been estimated that nearly four-fifths of Germany's entire aircraft production resources fell into Soviet hands. Luftwaffe production centres in Poland, East Russia, East Germany, Hungary, Czechoslovakia and Roumania were taken over by the Soviet authorities. A Soviet eyewitness account of the take-over of the Junkers plant at Dessau gives some measure of the wealth and importance of the German aeronautical booty. It has been given by a member of their Commissariat of Aviation. It says: "There were factories, branches of Junkers and other firms, which had not been dismantled by the Americans and which were still working. In one small place on the banks of the Mulde, we discovered a former paper factory, which was still engaged on the assembly of jet engines at that time. Two storeys of the building were filled with crates containing new turbines in perfect order. The same sort of thing happened in the Siebel plant at Halle, the Junkers engine factory at Bernberg and the Heinkel factory at Oranienburg." (For further details see V. L. Sokolov: *Soviet Use of German Science and Technology 1945-1946* published in New York in 1955.)

The above examples are not difficult to multiply. In Prague the Soviet obtained not only many flyable samples of the twin-jet Messerschmitt 262, which was being assembled in that city, but also some of the Walter rocket engines, capable of giving some 4,400 lb. thrust. At that time this engine powered the Messerschmitt 163 rocket fighter. At the Heinkel plant at Warnemunde they found the Heinkel 162 jet fighter in production. They also came across the design of a more powerful Heinkel jet engine, the S-011, and also the blue-print of an advanced rocket-powered Heinkel fighter aircraft. At Spandau near Berlin, they received further samples of the Bayrische Motoren Werke's BMW 003 co-axial jet which was to power some of the first Soviet jet fighters. At one of the Junker plants they found full details of the experimental four-jet bomber the Ju 287K, which had already been test-flown.

The aircraft component plants of Siemens and Askania yielded to the Soviet authorities important radar equipment which is discussed in the chapter on "Strategic Air Defence".

No wonder, then, that specialists and delegations from the Soviet Commissariat for the Aviation Industry, from TSAGI, the Central Aerodynamics and Hydrodynamics Institute in Moscow, from VIAM, the all Union Institute of Aviation materials and from NISO, the Soviet Scientific Research Institute for Aircraft Equipment, were all sent to scour the resources of Germany's most modern aircraft factories and research institutes, the latter including the Deutsche Versuchsanstalt für Luftfahrt (DVL), the Kaiser Wilhelm Institute, where research on the atomic bomb was being carried out, and the Luftwaffe and German Army test centres at Peenemunde and Heidelage, where long-range rocket development was in 1945 more advanced than anywhere else in the world.

In the final analysis, the Soviet rocket and missile legacies, both the offensive and defensive weapons, were the most important in the long catalogue of air weapons which they inherited from Germany first because Germany was well ahead of any other power in the world in this field in 1945; and also because we are passing into the military age in which manned bombers and fighters are being gradually replaced (though not entirely replaced) by guided rockets and missiles. Peenemunde, the rocket research institute on the Baltic Sea, was established by the German Army in 1937. Like Stalingrad, Malta and Alamein its name has a permanent place in modern military history. It reached the headlines of the Western press after the night of August 17th-18th, 1943, when it was successfully raided by R.A.F. bombers. Its first operational product was the pilotless V 1 flying bomb with a warhead of about a ton, which the Soviet acquired. Despite its novelty, the V 1 has, however, no important rôle in the contemporary range of Soviet air and rocket weapons, for its speed of about 360 m.p.h. rendered it obsolescent in the post-war period. Not so the V 2 long-range rocket, which first penetrated into the London area on a September evening in 1944.

While it is true that some of the top German V 2 rocket specialists made their way to the West, among them Dr. Wernher von Braun and General Dorburgher who have made notable contributions to the contemporary development of United States' long-range missiles, the Soviet rounded up nearly 200 rocket specialists from Peenemunde in 1945-46, as well as several hundred other German workers and technicians manufacturing the V 2 rocket or its power plant. Then there were the missile-tracking and guiding specialists from the German firms of Askania, Lorenz, Siemens and Telefunken, not to mention almost the entire personnel and equipment of one of the German V 2 rocket test stations. Many of these German rocket specialists were taken to Ostashkov, north-west of Moscow. Others went to Kuibischev, some worked at Khimki on the post-war Soviet long-range rockets under the general direction of Professor Artakianov of the Soviet Academy of Science. Some of the Germans at Kuibischev also worked under the direction of Soviet rocket specialists. Others who were chemists and fuel specialists went to work on these problems at other rocket development institutes near Moscow and Leningrad.

In these days of well-tested rocket missiles both for offensive and defensive operations, it is easy to underestimate the value of the German pioneer work in these fields. A new military era began in

July 1942 when the first V 2 long-range rocket was tested. It rose three feet, exploded violently and destroyed its own testing station. But by the end of the Second World War over 4,000 V 2s had been successfully fired against London and Antwerp, and German scientific engineering genius had presented both to the East and the West a basically revolutionary air weapon which speeded up the recent advent of earth satellites and space travel and the potential military use of intercontinental long-range rockets in the nuclear age, rockets which in the long run may in fact do more to keep world peace than any other effort, human or technical.

The V 2 German rocket was a primitive short-range missile compared with the sophistications and strategic potential of today's-ICBM and IRBM missiles which are being manufactured on both sides of the Iron Curtain. But it was an important weapon which gave an accelerated start to the post-war Soviet programme of long-range rocket development. It was highly mobile and required no launching tower of any kind. With a take-off weight of some 12½ tons and a rocket motor developing some 60,000 lb. thrust at sea level, it reached a maximum speed of about 3,600 m.p.h. at a height of 22-23 miles. Its oxygen and alcohol fuel supplies could be cut off by radio ground control or by a radio switch in the integrating accelerometer. With a range of some 200 miles and a warhead of less than a ton of explosive the V 2 would have virtually no place in the present-day series of long-range rockets. Dr. Ley in his *Rockets and Space Travel* (Chapman & Hall, page 227) has described the German V 2 as "not quite a finished development which was ready for military purposes to just about the same extent as airplanes were in 1914". The development of long-range rockets both in the USSR and the U.S.A. in the fifteen years that have elapsed since the first operational firing of the German V 2s have borne out this contention.

But more important than the V 2 weapons themselves were such top rocket German scientists as Professor von Bock and Dr. Wilhelm Tellman. Tellman escaped from the USSR at the end of 1949 via Greece to tell how under the direction of a Soviet rocket scientist he and his colleagues from Peenemunde and Heiderlage helped to develop the Soviet improvement of the German V 2 rocket (believed to be the T 1 rocket) at experimental and production rocket centres in Leningrad, Tomsk, Kazan and Irkutsk. The German effort in the Second World War had shown the Kremlin that rocket motors of great size could be mass-produced, and that large long-range military rockets could be controlled sufficiently to make them reliable military weapons. Even if they lacked accuracy at this stage the Soviet General Staff appreciated that the V 2 was the forerunner of many future rocket weapons in the atomic age.

But it was not the V 2, known to Germans as the A 4, which interested Stalin most, but the A 9 and A 10, German rocket projects which in 1945 were only in the design and blue-print stage. The A 9/A 10 combination is believed to be the final German long-range rocket which was conceived, though not test-fired, in World War II. It was two-step, with a proposed weight of some 85 tons and intended to be transatlantic in range. It was probably intended to be fired at New York. It was this rocket combination to which Stalin referred at a post-war 1946 Soviet staff conference during which he threatened President Harry Truman with its use, though in fact it took more than a further decade before the USSR test fired its first intercontinental ballistic missile and by then Stalin was dead. In this new two-stage long-range rocket the A 10 section was to have been the booster rocket, which was probably to have been jettisoned by parachute at a height of about 80,000 feet when the main A 9 rocket would be fired and reach a maximum velocity of over 6,000 miles an hour and a height of perhaps 150 miles. It would cross the Atlantic in a gradual stepping descent. As it reached the denser atmospheres surrounding the earth its wings would be so brought into play, that the rocket got lift again, just as a flat stone hurled across the surface of a pond does. The Germans were working on an atomic warhead for this A 9/A 10 project to which Dr. Sänger contributed the design blue-print based on the mathematical ideas of Professor Herman Oberth. But they never got as far as testing the practicability of this two-stage rocket weapon. But they certainly provided personnel, ideas and stimulation for the Soviet development of a two-stage atomic warhead rocket in the 1950s. Among the top German scientists who went to the USSR were Professors Vollmer and Herz who had been specialising in nuclear physics at the University of Berlin, which they left for a nuclear physics institute in the Crimea.

The German legacy for Soviet long-range rocket development was perhaps in the long run the most important war booty which the Kremlin acquired. The long-range bomber inheritance from Luftwaffe resources was also of technical importance in the jet age that was then on its way in; but in general it was less decisive in its influence on the acceleration of the Soviet long-range bombardment programme during the decade that followed the Second World War. Although a prototype of the four-jet Junkers 287 bomber was secured, it clearly did not provide much help to Soviet heavy bomber designers such as Ilyushin, Myasischev and Tupolev, who in the middle and late 1940s and early 1950s were working on the problem of designing a 600-m.p.h.-jet bomber worthy of the new Soviet atomic bombs. The fact is that the Soviet were slow to develop long-range jet bombers in the post-war period. It could be because they overestimated the number of years required to develop the long range rockets for which the bombers were to be in part a substitute. A further factor was the Soviet inability to produce a really powerful jet engine of their own design in the first years after the war. Certainly the harvest of German equipment and engineers was of great importance to them, in the modernisation and improvement of their bomber production equipment. From one of the Junkers 88 factories they acquired two of the world's largest hydraulic presses, one with a capacity of 15,000 tons and the other of over 30,000 tons. These had been used to produce wing spars in a single operation. Over ten years later, in 1957, the Soviet proudly displayed presses of these types at an agricultural and industrial exhibition in Moscow.

One group of German bomber aero-engine specialists were despatched in 1946 to Kuibischev where they worked in the design and development bureau of N. D. Kuznetsov, who to-day is well known for producing a turbo-prop engine of 12,500 shaft horse power, the largest of its type in the world, which powered the Soviet Bear four-engined bomber designed by Tupolev in the early 1950s. This was the first Soviet bomber to have genuine intercontinental range. It is now obsolescent and the Kuznetsov turbo-props are used to power the TU 114 four-jet air liner, the largest passenger-carrying aircraft in the world, though because of its turbo-prop engines clearly not the fastest. How much of the design work for the Bear and TU 114 engines must be credited to German Junkers engineers is hard to determine; but it is certain that they made a valuable contribution which speeded up Soviet jet engine development. Other German aero-engineers, for instance Rudolph Rental, a senior Messerschmitt engineer, Adolph Beta, a German swept-wing specialist and Siegfried Gunther, a senior designer, who had had experience with both the

Heinkel and the Junkers firms, also helped to develop some of the post-war Soviet jet planes.

We have mentioned elsewhere the considerable German engine contribution to the early Soviet post-war jet fighters, such as the MIG 9 and the YAK 15. But the German contribution to the development of more important strategic air defence weapons such as Soviet ground-to-air and air-to-air guided missiles was even greater, especially as Germany was so far ahead of the rest of the world in designing and developing these weapons, though not in using them operationally.

One of the most interesting of the German missiles was the Bachem BP 20 Natter. Dr. Bachem's design for a piloted missile interceptor had been accepted by the Luftwaffe in 1944 in preference to those submitted by the firms of Heinkel, Junker and Messerschmitt. It had a top speed of over 600 m.p.h. and a rate of climb of 37,000 feet per second. It was to fire 24 × 13 rockets in one salvo against raiding bombers.

The BP20 Natter, the German Messerschmitt 163 rocket fighter, the German DFX 346 rocket project of the Siebel Company and the Heinkel rocket powered P 1077 and 1080 (the latter was fitted with two ram jets) provided a basis for the development of Yakovlev's experimental rocket fighter of the late 1940s and the Sukhoi supersonic rocket-cum-jet fighters of to-day.

However, neither in the Soviet Air Force nor in any of the Western air forces was a piloted rocket fighter or missile to play a major rôle in the new post-war weapons of strategic air defence. The unmanned radar-controlled ground-to-air rocket missile was to be the key to air defence in the new atomic air age. In this field of weapons the Soviet legacy from Wehrmacht engineering and scientific resources would be difficult to exaggerate. The most advanced of the German ground-to-air guided weapons, which was still untested in battle was the Schmetterling (Butterfly) designed under the aegis of Professor Wagner by the Junkers firm. The Schmetterling looked rather like a small aircraft with a wing-span of just over six feet and a length of some thirteen feet. Its two rockets units gave it a maximum speed of 620 m.p.h., a range of about 20 miles and a ceiling of 50,000 feet. This ground-launched missile was guided by radar. At the time of the German collapse in the summer of 1945 the first production models were ready to go into action.

In addition to the Schmetterling, the Soviet acquired one or two different types of the experimental Wasserfall (Waterfall); this was a larger 24-foot ground-to-air guided rocket with either a proximity fuse or a radio fuse which could be exploded from the ground. It had four stubby wings near the centre of the torpedo-shaped rocket. German records suggest that it had fired successfully in about one case in four. Finally in the ground-to-air missile category was the Rheintochter (Rhinemaiden) which was nearly twelve feet long but much more slender than either the Schmetterling or Wasserfall. It had six swept-back wings at the tail end and four control surfaces near the nose in which was a proximity fuse. It had radar control of the beam-riding type, but its range and service ceiling were not considered satisfactory by the Wehrmacht in 1945.

In the air-to-air guided-missile category, the Soviet acquired the experimental Feuerlilie (Firelily) which was really a high-speed rocket-powered small aircraft, designed by Dr. Eugene Sänger, better known for the stratospheric A 9/A 10 rocket plane project mentioned earlier in this chapter. But more important for the Soviet was the Henschal 298 air-to-air missile which was radar-guided, had a range of 2-3 miles and was about six and a half feet long.

German air specialists will know that the list of engines, aircraft, missiles and material already presented in this chapter is far from complete. It has not for instance included two fully tested and operated guided bombs used against British and U.S. shipping in the Atlantic and the Mediterranean in the 1943 and 1944 periods: these were the Hs 293 winged missile weighing some 1,700 lb. and measuring about eleven and a half feet long and the Fx 1400 rocket bomb which was more powerful but much less used in action. Despite some operational successes mainly against U.S. and British ships at the Salerno landings of 1943 and against the surrendering Italian fleet which was trying to make for the shelter of allied ports later that year, the radio guidance mechanism of these two bombs, usually installed in a Ju 88 or Dornier 217 bomber, was not entirely reliable. But in any event these subsonic guided weapons had no important place in the post-war armoury of air weapons in the USSR.

Other German air legacies included the parachute equipment and transport aircraft and gliders from the Luftwaffe school at Stendhal, the designs for Kurt Tank's ram-jet fighter and a synthetic fuel plant at Leipzig.

But perhaps the most important of the German air legacies of World War II were the lessons in tactics learnt by the Soviet in air combat against the Luftwaffe and the realisation that their pre-war standards of training in blind flying and the use of radio equipment were totally inadequate. The stringent measures that the Soviet authorities took to remedy these deficiences in the last year of the war and in the immediate post-war period are described in the next chapter ("Training and Personnel"). They were undoubtedly prompted in part by the ready Soviet acknowledgement of the much superior standards in these fields which they encountered in their German opponents. The raising of the poor standards of Soviet night-fighting and night-bombing in the late 1940s was due to some extent to the immediacy of the German example in these operations witnessed in four years of air combat.

It was in October 1946 and in the following months that the Soviet authorities swooped on hundreds of thousands of skilled German technicians and specialists, put them into teams and then on arrival in the USSR settled them into the machinery of the Soviet aircraft production industry. In addition many thousands of skilled Germans who remained behind in the Soviet zone of Germany quickly resumed work on making radar equipment and guided missile parts for the post-war modernisation of the Soviet Air Force. Some German estimates put the figure for the 1946 exodus to USSR factories at 300,000.

Only the combined documented testimony of those thousands of Germans who have returned to Germany after working for a decade in Soviet plants could tell the full and final story of the great German air legacy which in the jet, atomic weapons and rocket missile age, did as much as anything to enable the USSR to close nearly all the technical air gaps between herself and the West in little more than a decade after the end of the Second World War.

CHAPTER 5

TRAINING AND PERSONNEL

THE standards of air and technical training in the USSR have progressed greatly in the last decade, if only because the technical equipment and educational standards in both air and rocket training centres have improved. To-day and in the foreseeable future there is no reason to suppose that the skills and disciplines in Soviet air training schools, institutes and colleges are or will be inferior to those in the West. In an age in which long-range radar, defensive and offensive guided rocket weapons, helicopters and supersonic jet and rocket-engined aircraft dominate the military scene, personnel and training standards in the USSR are high in all these fields.

This is due not so much to the military standards of air and technical training, but to the great advances made in Soviet technical education and in technological manufacturing. Various figures and facts have been quoted to show the high status of the scientist and the engineer in the USSR and of the facilities which exist for producing large numbers of high-grade technicians. It is said (see, for instance, *Report on the Atom* by Gordon Dean, Chairman of the U.S. Atomic Energy Commission) that of some 200,000 Soviet graduates who qualify annually, about one-half are scientific or technical students. The total number of engineering graduates is estimated variously at 30-50,000 a year. There are said to be over 3,500 good technical schools in the USSR. Many leading scientific personalities in the West such as Professor P. M. Blackett, British Nobel Prize Winner in Physics, Dr. Bronowski and others are of the view that the USSR is turning out more technicians than the Western world. This is the result of a forty-year policy which has made the scientist in the USSR like the Party man, one of the chief household gods. In an age in which the chemist, the physicist and the engineer are replacing to some extent the aircrew member, the tactician and the strategist, the Soviet Union is helping to ensure high standards of schooling in its air force by having high standards of technical and scientific training in schools and universities. And this will become increasingly important in the next decade or so, as rockets become more accurate and tend to replace manned aircraft more and more. In an age in which rockets, atomic energy and electronics dominate the military weapons situation, the training and education received at Soviet technical schools, technical institutes and universities are at least as important in assessing Soviet Air Force personnel standards as the schooling given in the Soviet Air Force Training Schools.

The two cannot of course be divorced. In nearly all aspects of Soviet life, the link between civil and military is closer than in the West. Many of the leading Soviet politicians, including Khrushchev, have held military ranks in the political police. Soviet aircraft designers such as Tupolev and Ilyushin are generals, and the civil and military air transport resources of the USSR are much more intimately linked than in the U.S.A. or in Britain. The same is true of the university system of education in the USSR. And because Soviet air and rocket forces draw so heavily on Soviet universities, something of the general nature of university schooling and of its links with the military needs to be said.

It has been estimated that during the last forty years some four million students have graduated in science or engineering from Soviet universities and high technical institutes, though many thousands of the best graduates were killed during the Second World War. The teaching given is fairly uniform and there is a system of dual control which is said to help maintain high standards. The Ministry for Higher Education is responsible for drafting the syllabus and appointing professors and teaching staff, while other ministries (e.g. for aircraft production) contribute both to the cost and the policy of scientific research. The students follow the same basic syllabus and lectures in all the universities of the USSR. All lectures are compulsory and there are usually about 35 one-hour lectures per week. This does not include the practical work in the laboratory or on the bench. Such is the scheme for the first three years of study during which there tends to be a larger number of written and practical test papers than in comparable courses in the West. During the fourth, fifth and sixth years at the university or high technical institute there is a greater degree of specialisation in, for instance, nuclear physics, electronics or astronautics than in the majority of Western universities. There are more professors holding specialised chairs which the Russians call a *kafedra*. Moscow State University alone has over 150 specialised professors of this kind.

Visiting Western scientists and a number of escaped Soviet scientists and technicians have reported that standards of experiment and criticism are as high as in the West. The Lysenko theory which achieved great publicity in the West was an exception which did not prove the rule. Roughly ten per cent of lecture time is devoted to political indoctrination; but the sessions on Marx-Leninism, political economy, dialectical materialism and the history of the Communist Party do not seem to have any adverse effect on the levels of study or research. As one Soviet graduate put it, "I imagine that these lectures had no more effect on my fellow scientists at Kiev University than they had on me". They are part of a national catechism and are accepted as inevitable. Some of those who should know think that the Soviet State Examinations in science and engineering demand a high standard, sometimes higher than in Western universities. They nearly always ask the student to write a thesis, but he is not awarded a degree even at the graduate stage. If successful in the State Examination, the student can go into industry, teach, or do a post-graduate two-year course called *aspirantora*. At the end of this period he or she can then sit for the degree of "Candidate of Science". The next highest academic qualification is Doctor of Science, which varies in length according to the research undertaken. The Doctors of Science become members of the Soviet Academy of Science or hold chairs at a university. Many of them lecture at the officer engineering schools and air academies of the USSR and help to draft the technical syllabus for the Soviet military and air technical institutes.

The links between the civilian universities and higher technical institutes, the Soviet Defence Ministry and the teaching schools and academies of the Soviet air and rocket forces are many. In fact, apart from appointing staff, Soviet civilian technical higher institutes are concerned with nearly all facets of advanced technical training in the Soviet Air Force. One of the facilities for creating such links is the much larger number of chairs of military science at Soviet universities than in the West. They are frequently, but not always, occupied by fairly senior military officers. Apart from lecturing on military problems they also run recruitment and training programmes of direct value to Soviet air and rocket forces outside the general framework of the Soviet Air Force. There is for instance a scheme run at Soviet universities called the VVP (Vyshaya Vnevoiskovaya Podgotovka), i.e. "training outside the services". The system is not new, for it was inaugurated in the 1930s. It has expanded in the post-war period to take in a much higher proportion of Soviet universities and higher technical training institutes. The idea behind the scheme is to catch up with university students who may be exempt from military service because they are going on after graduation to an important job in industrial science or engineering. Many of them are selected to be reserve navigational or engineering officers for the Soviet Air Force and are given training while at the university. After two years of training, mainly in vacation time, they are vetted by a so-called credentials committee and if found both competent and politically reliable become fully fledged reserve officers. The value of such a scheme need not be emphasised; it means that the Soviet Armed Forces have a stronger link with top-grade technical civilian graduates than exists in Western countries.

The number and scope of air academies in the Soviet armed forces have greatly increased in the post-war period. The oldest and best known are the Zhukovsky Air Force Engineering Academy and TSAGI, the Central Institute for Aerodynamic Development. The Mozhaisky Air Force Engineering Academy at Leningrad and the large Orjohikidze Aviation Institute in Moscow also go back to pre-war days. These are intended mainly for experienced commissioned officers who are likely to become members of the General Staff or staff officers of air armies. But there are also smaller air force engineering institutes specialising more and more in rockets and guidance mechanisms at Kiev, Kharkov, Voronezh, Odessa, Komsomolsk, Rybinsk, Kazan and other main cities in the USSR. The Air Force engineering college at Riga lays emphasis on aero-engines and electronics. At the Kuibischev Air Academy, east of Moscow, there is specialisation in airfield construction and maintenance, as well as courses on jet and rocket engines. The air Academy at Monino like the Zhukovsky Academy in Moscow tends to give more courses in strategy, tactics and administration for Soviet Air Force officers who are likely to have senior appointments in air divisions, air corps or air armies.

Clearly in the post-war period expansion in electronics training has been inevitable. A signals air academy has been set up in Moscow which trains officers for responsible posts in the early warning system of the USSR or as radar or signals officers in one of the local sectors or areas of the PVO strategic air defence system. Other good radar and signals training schools, mainly for N.C.O.s and other ranks, are to be found in Novosibirsk, Khabarovsk, Riga, Kiev and Leningrad. The first course at one of these electronics training schools lasts about twelve months. Students are given thorough instruction in all the major aspects of servicing and using Soviet wireless and radar equipment. At first they learn to service and use ordinary wireless and V.H.F. sets such as the RS 16M and the RS 6R sets, which take the normal air-to-air, air-to-ground and ground-to-air traffic. Then they go on to various forms of radar equipment such as the Soviet ground-controlled approach for landing at night in bad weather, the airborne radar for night-fighting, the air-to-air defensive guided-missile systems, and the radar gunsights used for instance on the MIG 19 and 21 fighters. Then there is the study of the radar guidance systems of such ground-to-air missiles as the T 6 and T 8.

It is of course difficult to say or even guess how effective the overall standards of Soviet electronic training are compared with those in the West. Clearly Soviet ability to use early warning radar effectively has been demonstrated practically in the Korean War and in the interceptions of unarmed British and U.S. aircraft in various parts of the world. The Soviet jamming of Western broadcasts is technically better than the German efforts in this field in World War II. Therefore it would be a mistake to suppose that the Soviet will not or cannot in general reach training and operational standards comparable with those pertaining in Britain and the U.S.A. Nevertheless the Soviet are aware that all radio counter-measures and all early warning systems are likely to have local failures. This is so to speak an important part of the deterrent, for it makes any part of the Soviet Union liable to be wide open to attack from the West.

The use of better and more modern electronic equipment in Soviet air training schools in the last few years has raised the entire standards of effectiveness of the operational training courses for fighter and bomber units. As more up-to-date V.H.F. equipment is installed, fighter tactics can be improved more easily and flying disciplines more readily established. The use of ground-controlled interception and airborne radar has made the night-fighter operational and training on the Ilyushin 28 and Yakovlev 25 twin-jet planes more purposeful. In particular the navigational night- and blind-flying training in the Soviet D.A. long-range bomber force has benefited greatly from the introduction of radar and radio navigational aids.

The Soviet Air Force General Staff is clearly alert to the importance of maintaining high standards of training in all sections of the Air Force. Despite the intoxication of its very successful operations in the last two years of the Second World War its air leaders were very conscious of the fact that in many respects, e.g. tactics and night and blind flying, many Soviet air units were at that time well behind the operational flying standards achieved in the U.S.A.A.F., the Royal Air Force and the Luftwaffe. The operations of the Soviet long-range command had been generally ineffective even when opposition was weak. Day-fighter escort in general in the Soviet Air Force did not, for example, reach the standards set in the French and British air forces. The Soviet had had practical examples of this when British flown Hurricanes escorted Pe 2s on the Murmansk front in 1941-42 and French flown Yakovlev fighters escorted Soviet bombers in the 1943-45 drives for Poland and East Prussia. Indeed a group of some 100 Soviet pilots flying escort in Soviet tactical air regiments on these fronts were sent back from front-line duty to be retrained at Tashkent and Sverdlovsk. There are reported to have been similar instances on other fronts in the last year of the war.

When peace came in the summer of 1945 there were two broad effects resulting from the inadequate training standards. First, there was a substantial reduction in front-line operational strength from over 20,000 in the summer of 1945 to about 15,000 at the beginning of 1947. It was felt to be more practical to have a smaller but more efficient air force in which higher training and personnel standards could be reached. Thus three of the six air corps based in Germany at the end of 1945 were disbanded. Severe reductions were made in the number of air regiments in the Air Army under General Krassovski's command based in Austria and Hungary. And in Roumania two of the four air corps in General Kamanin's 3rd Air Army lost their operational status. The Air Armies in White Russia and the Ukraine made similar reductions. It was clear that at the end of the Second World War, the new Soviet Air Force Commander-in-Chief, Chief Marshal of Aviation Novikov, as he was to become, had absorbed one of its major lessons. So often in the Italian Air Force and in the Japanese and German air forces of 1943-45, partially trained aircrews had been rushed into front-line service with indifferent results. The coming jet, rocket and atomic age was to call for even higher training standards than before. Novikov made his very capable deputy, Colonel-General Sudetz, who was later to command the Soviet long-range jet bomber force, responsible for the expansion and modernisation of the Soviet air training scheme in the late 1940s. Marshal Novikov and his senior air colleagues, such as air generals Sudetz, Rudienko and Golikov, gave a series of lectures at staff colleges and air academies, and wrote articles in the military press emphasising the importance of increasing the amount of flying practice, improving training equipment and reducing the theoretical lectures to the minimum. While clearly it is absurd to generalise about training standards in any air force it would appear that in the last decade Soviet air schooling has virtually closed the gap which existed between its training standards and those of the British, German and U.S. Air Forces at the end of the Second World War.

One of the first major measures in the new post-war training scheme was the setting up of new air training schools which concentrated on night- and blind-flying training. In the 1947-48 period training centres of this kind were established near Kursk, Irkutsk, Cheliabinsk, Gorki, Michurinsk, Tashkent and other centres. And instead of the commandants of flying training schools being operational commanders who had not been very successful under combat conditions, the new post-war chiefs of the blind-flying schools were usually successful commanders of air regiments and air divisions who had also graduated well at courses at the Soviet air academies.

Many of the basic features of present-day Soviet air and technical training go back to the 1930s. It was then for instance that Soviet youth were encouraged from early childhood to be air-minded and technically minded. The Osoaviakhim organisation of that period has now been replaced by an organisation called Dosaaf, a merger of three previous post-war institutions, Dosav, the Air Club, Dosarm, the Army Club, and Dosflot, the Navy Club. This merger of youth instructional military clubs, is a secondary but perhaps important example of how the USSR can go further than the West in combining the elements of the three armed services at all levels. The flying clubs of the post-war Dosaaf organisation are an important recruitment pool from which the Soviet air squadrons have drawn many of their best pilots and they are a fundamental element in the Soviet air training system. Before World War II there were over 100,000 aspiring pilots in the Osoaviakhim clubs undergoing some form of aircraft or flying instruction and it is a reasonable assumption that this figure has been maintained or exceeded in the post-war Dosaaf clubs. There is now, of course, even at this elementary stage, greater emphasis on electronics and rockets. The training also includes parachute jumping, at first from the static towers which are ubiquitous in all main cities of the USSR and later from aircraft such as the Antonov 2 biplane transport; and there is also elementary flying training instruction with such planes as the YAK 18 machine.

The Dosaaf clubs are not entirely concerned of course with military instruction. Their main attraction for Soviet youth is the boxing, athletics, camping, literary and other activities which include the making of models of aircraft and rockets. The Soviet authorities are justly proud of the rôle the Osoaviakhim clubs played in World War II, when some of their aero clubs became flying training schools for the Soviet Air Force. In addition many club members were able to graduate as pilots much more quickly than other Soviet personnel under training and take a leading part in front-line duties, e.g. in the Stalingrad air operations.

The basic air training scheme in the Soviet Air Force is not vastly different from that in any other modern air force. It divides broadly speaking into *ab initio* elementary training, the intermediate stage and the advanced operational flying, most of which in peacetime takes place when the pilot or aircrew member has passed out and been posted to an operational air regiment. The normal age for intake into the Air Force is about seventeen, though advanced technical students from universities and similar higher educational institutes come into the signals and rocket units at a later stage, sometimes up to the age of twenty to twenty-five. The Dosaaf clubs are becoming the main source of recruitment for flying personnel, but the Air Force entrance examination at seventeen does not call for any special knowledge of aeronautics, though the subject is included in the papers. Both the medical and general educational standards for entry have been raised in the jet and rocket engine weapons age. Candidates, as previously, have to satisfy the examining commission about the political reliability of themselves and their families and the Konsomol, MVD and KGB organisations all take part in these security checks which are part and parcel of general life in the USSR. They do not seem to affect Soviet achievements or discipline to the extent that many writers have depicted.

As in pre-war days, the aircrew training course at Soviet Air Force schools takes about two years to complete, though the length naturally depends on the type of air operations for which the training prepares. Long-range strategic bomber training is clearly a longer process than training for the tactical support operations of the so-called Soviet "Front Air Armies". The elementary *ab initio* training usually lasts from nine to twelve months and the standards of the past decade have been much higher than ever before, as the number of pilot instructors with combat experience has been much greater. However the first part of the course does not provide much flying experience. It includes a considerable amount of drill, guard duties, foot-slogging, military discipline and political indoctrination giving the latest party line on topical events. In the first year, however, courses lasting up to three months are given in the study of radio communications and radar and there are lectures on aircraft construction and maintenance, aero-engine design, gunnery and navigation, on rocket design, propellants and guidance and also an introductory course to Soviet Air Force flight manuals. Until an Air Force pilot candidate has passed the examinations in all these subjects, he cannot go on to the flying training courses themselves. In World War II there were some 150 flying training schools in the Soviet Union, but the number has been increased since the war to over 200. Part of this increase was to cater for many thousands of Chinese, Czech, Polish, Hungarian, Roumanian and Albanian pilots, as well as pilots from Egypt, Syria and Afghanistan. Only about a quarter of of the schools are concerned with elementary flying training. More important are the advanced training schools of which there are some fifty for fighter training. The emphasis at these operational training schools is more and more on strategic air defence and less on tactical support. They include expanded training schemes at Tashkent, near Moscow, Odessa, Grozny in the North Caucasus, Kizilarvat in Central Asia, Kuznetsk on the Volga, at Kustanai in Kazakstan and near Sverdlovsk.

There has been modernisation of most of the aircraft and equipment at all stages of Soviet flying training in the past few years and this process is bound to continue. The long service UT 2 (U stands for instruction and T for training) designed by Yakovlev in 1935 was replaced two or three years ago by a more modern machine, the YAK 18. It can land and take off in less than 300 yards and so is suitable for the relatively small airfields which the elementary flying training schools use. It is powered by a 160 h.p. engine giving it a top speed of 150-160 m.p.h. and a landing speed of about 50 m.p.h. There have been a number of reports that both Yakolev and Mikoyan are designing elementary and intermediate jet trainer machines fitted with engines of about 1,000 lb. thrust and so the YAK 18 may well be phased out of elementary flying training schools in the next few years; but, for the moment, the YAK 18 and its immediate successor, the YAK 20 of similar design and performance, predominate. The first twenty to thirty hours of training are dual-control flights with the instructor assessing the pilot potential of the new pupil. Then come between twenty and thirty hours of solo flying, beginning with the usual circuits and bumps and going on to long-distance cross-country flying.

The most generally used intermediate trainers are the Yakovlev 11 and 14. The former aircraft first came into general use in 1947. It is fitted with a Shvetzov 650 h.p. engine and in 1950 set up a world record for aircraft in this class by achieving an average speed of 274 m.p.h. in closed circuit competition flying conditions. Like the YAK 11 it has a range of 500-600 miles. Both the YAK 11 and 14 are likely to be replaced by jet trainers in the early 1960s, in fact the process has already begun. The intermediate stage of flying training lasts some three months, in which a further thirty hours solo flying is carried out. There are two important features at this stage. First the beginning of instrument and blind-flying training, and secondly the important decision whether the pilot aspirant is to specialise as a fighter, bomber or reconnaissance pilot, whether he is to transfer to a navigator specialist training school or if he is fit only for ground duties. In the latter case he is frequently sent to a Soviet Air Force technical school specialising in maintenance. Radio and radar personnel, rocket specialists and gunners are rarely recruited from the failures of the flying training schools. They usually come from the better-educated intake who at the age of seventeen or so fail to pass the increasingly stiff medical examination for pilots and aircrew.

It is in the advanced operational training stage, when the Soviet airman specialises as a bomber, fighter reconnaissance or transport aircraft pilot or crew member, that the training schedule varies most in length. Fighter pilots intended for the tactical air regiments go into front-line squadrons after a further seventy to eighty hours of operational schooling on the trainer version of the YAK 15, the YAK 17, the MIG 15 and the MIG 17. They concentrate on formation flying and close escort work as well as instruction in aerobatics. They have special lectures on army co-operation and close support operations. Unlike the war-time period, there is also a special course in blind and instrument flying, and quite often the best pilots are transferred to the longer operational fighter training for the strategic defence air regiments of the PVO command. Until a year or so ago there were still some obsolescent non-jet fighters such as the MIG 3 and the Lavochkin 5 used for advanced fighter training, but both these and the YAK 15 and YAK 17 jets are likely to be out of use soon. The MIG 15 and MIG 17 and perhaps a trainer version of the supersonic MIG 19 will probably be the three aircraft in general use for advanced fighter training.

Those pilots, usually of a higher calibre, who are selected as possible material for General Klimov's strategic defence fighter regiments usually do about 100 further hours of fighter training after qualifying at the intermediate stage of training. They go through the same type of schooling in aerobatics and formation flying as do the tactical fighter pilots, but they also have courses on the use of airborne radar and do flying exercises with ground control interception units. In addition their training on instrument and blind flying is more intensive. Their combat tactics training is also more extensive. Apart from the usual flying on the MIG 15 and MIG 17 trainer planes, some of the pilots are trained in all-weather flying on the Ilyushin 28 twin-jet and on the trainer version of the YAK 25 subsonic twin-jet aircraft.

The longest operational training courses are for the long-range jet bomber crews and it has recently been raised to 150-200 hours' flying, of which a high proportion consists of instrument and blind flying using the latest Soviet airborne long-range radar navigational aids. Some of the early bomber training is still done on the wartime twin-engined PE 2, TU 2 and IL 4 aircraft, but fairly soon the training crew goes over to long-range flying of 1,000 miles or more on the four-engined TU 4, the twin-jet TU 16 and the four-engined turbo-prop TU 20. If the pilot or navigator fails the long-distance flying and bomb-aiming tests, which include operational training tests over Arctic territories, he may be relegated to an air transport unit. If this happens, his training for an air transport regiment posting will vary from twenty to thirty hours to over one hundred depending on whether the pilot is intended for routine cargo carrying flights or for specialist work with the Soviet parachute divisions or the Arctic Air Transport Command.

However, the final edge on Soviet aircrew training standards is put on in the air regiments flying from their peacetime operational bases. The tactical air regiments of fighters and fighter bombers train in annual manœuvres with the Red Army. It is then that the latest procedures for speeding up air support for ground forces are tried out, that visual ground signal procedures, so important in the age of atomic weapons, are polished up, and the skills and mobility of the air squadrons are tested. Then of course there are the flying conversion courses from the MIG 15 and the MIG 17 to the YAK 25 as well as to the new supersonic Ilyushin tactical support bombers which are now coming into service.

The long-range bomber units of the ADD under General Sudetz do up to 5,000 mile training flights, especially from Arctic bases, during which they do bomb-aiming tests, long-range navigational training and flight re-fuelling exercises. The air transport regiments are so to speak in permanent flying training, but with the recent increase in the availability of long-range jet transport aircraft there is much conversion flying to the four-jet prop IL 18 and the AN 10 going on. In addition air transport regiments, many of them flying the YAK 24 and MIL 6 helicopters, take part in the annual Red Army manœuvres in which they drop troops, tanks and artillery and carry supplies. Among the hardest worked of the air regiments are probably the strategic fighter defence units of the PVO. Like the long-range bomber units, they carry out intensive operational training in instrument and blind-flying procedures and on co-operation with ground-to-air guided-missile batteries. This calls for precise navigation and co-operation with the ground control organisation as the zones of operation have to be most clearly delineated. In the uncertainties of radar identification, the PVO fighter units might well be shot down by the ground-to-air guided-missile batteries. This problem of course is not peculiar to the Soviet Air Force, but is now much more acute than in World War II, when it was only the general inaccuracy of anti-aircraft fire which prevented more fighter and bomber planes being shot down by friendly gunfire. The PVO fighter units are also engaged in testing the new airborne radar equipment and there is a considerable amount of conversion flying going on from the MIG 17 subsonic and MIG 19 supersonic planes to the faster MIG 21 and Sukhoi delta fighters now coming into service.

Incidentally it would not be surprising to learn that the Soviet Air Force have introduced a combined operational training scheme for tactical and strategic fighter units. This is because, as the number of defensive guided missiles increases, the fighter regiments may be required to do more escort and protection of transport and helicopter machines as well as helping to maintain local air superiority in a local war and in regional support of ground forces. On the other hand, the USSR can never quite know its full defensive fighter commitments. For example it might suddenly have to provide seriously for the metropolitan air defences of China, which at the moment are very thinly manned.

As the long-range missile potential of both Western and Soviet military forces grow, so will the emphases in operational training change. Clearly fighter training programmes will become less ambitious in the next few years, while the facilities for training defensive guided missile units will be expanded. In the bomber fields, both tactical and strategic, the same trend is likely. The emergence of Soviet ocean-going submarines as a major component in the long-range atomic attack potential of the USSR will produce an expansion in long-range naval air training, so that the techniques of co-operation with Soviet submarines can be improved. And as manned aircraft play a less prominent part in overall Soviet air potential, so the emphasis will be not on flying training as described in this chapter but on technical education about such things as rocket fuels and electronics and guidance systems. The airmen of the future will become more and more like the polytechnic student, and the development of scientific training in schools, universities, higher institutes and factories will be more important than producing a three-point landing or doing flick rolls and tight turns. But as far as can be foreseen there will always be a demand for Soviet fighter, bomber, reconnaissance and transport aircrews in the next few decades, as long as targets are mobile and fleeting. The limitations of aiming a rocket, even one which is one hundred per cent accurate, is that it must be fired from a fixed point to a fixed point. Later on more sophisticated rockets may be able to chase their quarries and hit mobile targets. Until then troops, tanks and mobile batteries of guns or rockets must be dealt with by manned aircraft. And so in part the Soviet air training scheme will remain basically what it became in the first decade that followed the Second World War.

CHAPTER 6

AIRBORNE FORCES

THE Soviet Union is fond of claiming inventions and innovations. Exaggerated national pride in a young revolutionary régime is understandable. It is surprising that the Soviet authorities have not in their propaganda more frequently claimed to have invented the notion of using parachutist and airborne troops. It is a generally recognised and a fully established military fact that the Red Army was the first to organise units of parachute troops and to conceive the idea of transporting large bodies of equipped infantry by air. This they did five years before the Germans; but for those with literary interests it should be added that the English sage Dr. Samuel Johnson had, in the eighteenth century, prophesied that the day would come when armies would capture whole cities by airborne assault. With the growing capabilities of modern long-range jet and other transport planes, who would deny a potential reality for Dr. Johnson's imaginative prediction?

Although on paper Soviet parachute units were formed in 1930 the first two years produced only modest beginnings. Voroshilov, then People's Commissar for Defence, spoke proudly to congresses and meetings in the USSR about the new air arm: "Parachuting is a phase in the field of aviation which has become a monopoly of the Soviet Union. Any surpassing of us is inconceivable," he declared. It is true that in the early 1930s other countries lagged behind the USSR in this military field. But by 1940 and 1941 the achievements of the German parachute formations in Holland, Belgium, Norway, Greece and Crete were certainly not to be surpassed by any Soviet airborne and parachute operations of the Second World War. Nor can it be said that Soviet airborne formations and parachutists achieved more than British and American formations in the conduct of airborne operations. It is the purpose of this chapter to explain the uncertain progress of Soviet forces in this field both in wartime and in the pre-war period, and to assess the present and future potential of Soviet airborne troops in the age of atomic weapons, rockets and jet aircraft.

The Soviet airborne forces developed at a leisurely pace in the 1930s. The first understrength brigade was formed in the Leningrad area in 1932 and consisted of some 450 parachutists, with about a dozen TB1 bomber and R5 reconnaissance aircraft adapted for parachute jumping. The first parachutes were of American manufacture. The jumping instructors came from Soviet Air Force units and the parachutists mainly from Red Army units. Among the early outstanding Soviet parachutists were Maschkovski, Minayev and Grochovski. In 1932-35, following the relative success of the Leningrad unit, Marshal Tukhachevsky encouraged the development of new airborne and parachutist brigades formed from troops in the military districts of Minsk, Kiev and the Far Eastern Maritime Provinces. The adaptation of the four-engined Tupolev TB 3 bomber as a transport aircraft in 1934 speeded up the training and dropping exercises and by 1935 the Soviet High Command felt that the new airborne arm was ready to show its paces before foreign military observers. In parachute manœuvres at Kiev in that year over 500 parachutists were dropped from a fleet of TB 3 transports, a further wave of these planes landed a battalion of airborne infantry and in the same manœuvres supplies and artillery were also landed by air. What the 1935 manœuvres did not reveal was the extent to which Soviet parachutists had for the past few years been training extensively on reconnaissance and sabotage exercises, that is for operations behind enemy lines. It was also in 1935 that a complete and fully equipped rifle division of 14,000 men was transported by air from Moscow to Vladivostok. This at the time was an air feat without precedent, a warning to the Japanese of the speed with which the Far Eastern Soviet Armies could be reinforced and a forerunner of the successful Soviet airborne reinforcement to the hard-pressed Stalingrad garrison at the end of 1942.

Encouraged by the success of the public demonstrations and manœuvres of 1935, the Soviet parachutists and airborne troops continued to expand and develop in the next two or three years. In 1936 there were more extensive demonstrations given in both White Russia and the Ukraine. In each district the local parachutist forces had been increased to a strength of two battalions of about 300 men, further training had been carried out in the dropping of machine guns

SUKHOI DELTA-WING (NATO CODE-NAME: *Fishpot*)

MIG 21 (NATO CODE-NAME: *Faceplate*)

SOVIET JET FIGHTERS

TU-20 LONG RANGE BOMBER (NATO CODE-NAME: *Bear*)

IL-18 MEDIUM RANGE TRANSPORT (NATO CODE-NAME: *Coot*)

SOVIET HEAVY AIRCRAFT

and light artillery and the introduction of a double parachute providing greater safety led to greater flexibility in the heights at which both weapons and troops could be dropped. At the main Ukraine manœuvres of 1936, before many foreign military attachés, 1,200 parachutists were dropped from TB 3 aircraft and 5,000 airborne troops were landed together with their equipment, machine guns and light artillery. In 1937 there were further public demonstrations and developments. Parachutists from the airborne brigade of the Moscow Military District showed that the quality of dropping was high and demonstrations of the parachuting of light vehicles, guns and armoured cars indicated further progress in this field. There were at the time a number of reports suggesting that the current Soviet Army tank, the T 37, had been successfully dropped by air from a TB 3 four-engined machine, but there is no subsequent record that this was ever attempted under combat conditions in World War II.

From 1937 until the outbreak of the war with Finland in the winter of 1939-40, there was an air of official diffidence about the development of Soviet airborne troops. The annual public manœuvres ceased. At Party conferences, the Soviet Commissar for Defence spoke proudly and in some detail of many Red Army developments, but made little or no mention of the parachute and airborne forces. None of these formations was sent to the civil war in Spain, though it would have been an excellent opportunity to test the guerrilla and sabotage techniques and tactics which had been developed in the previous five years. No one outside the Kremlin can know all the reasons for this apparent major change in Soviet military policy. However, some of the reasons for the slowing up in the development of Soviet parachutist and airborne forces in the late 1930s are apparent. In the Red Army purges of 1937 and 1938 Marshal Tukachevsky, the chief organising brain on the General Staff and perhaps the chief military sponsor of Soviet airborne forces, was liquidated. Many senior officers in the airborne troop units, including General Gorbatov, were imprisoned. Others who suffered in the purges at this time include such men as Tupolev, chief designer of long-range aircraft, who was arrested and later released, and General Khripin, an advocate of greater development of long-range bomber and transport aircraft for the Soviet armed forces.

More important than the purges were the difficulty in producing modern heavy transport and bomber planes and the importance of concentrating the aircraft industry on the modernising of the fighter, ground attack and medium bomber categories and of building the largest possible tactical air force to meet the possibility of a two-front war waged against the Soviet Union by both Japan and Germany. Tupolev had designed a successor to the four-engined TB 3—the TB 7, the plane which was to bring Molotov to London in 1942. It could have been suitably adapted for dropping parachute troops; but to produce it in any quantity would have absorbed too many man hours and slowed up the development and production of the new MIG, YAK and LAGG fighters. The TB 3 continued to give satisfaction as a troop carrier and parachutist aircraft, and was extensively used in this rôle in the Finnish War. But it was also in demand for the Soviet bomber units. Nor was the Kremlin alone in having to compromise between two operational requirements. For instance the British decision to put the major effort into the creation of a long-range bomber force in 1941-42 was at the expense of an adequate air transport command. While the results obtained from strategic bomber operations against Germany were important, the British air transport resources to support ground forces in the summer advances in France and Belgium in 1944 and the crucial airborne operation at Arnhem in that year were certainly inadequate.

However, despite the Soviet purges and the growing shortage of suitable transport planes, the Soviet airborne forces continued to expand. By 1938 there were four brigades each of over 2,000 parachutists training in the military districts of Leningrad, Kiev, White Russia and the Far East; and at the special paratroop schools of the Soviet Air Force more and more Red Army officers and men were undergoing the stiff intensive four to six months' training. Much of the preliminary training and interest in parachute jumping was developed in the para-military youth organisation Osoaviakhim founded in 1927. It encouraged young men and women in the Soviet Union to take up parachuting as a sport in preparation for army or air force service, as well as providing other primary defence training and opportunities for athletic, literary, social and political pursuits. The Osoaviakhim organisation provided many thousands of pilot candidates for the air forces and encouraged many thousands of Red Army infantrymen to become parachutists, just as its post-war counterpart Dosaav does to-day.

At the Soviet Air Force parachute schools, there was a dual process of instruction and toughening. Day marches of up to 50 miles had to be completed by nearly all parachutists, all arms from rifles to heavy tanks had to be handled and serviced by most of them. There were courses in map-reading, using and repairing signals equipment and in engineering, artillery and command problems at the section, platoon, company and battalion levels. The emphasis varied according to the type of parachute unit for which the candidate was intended. Most parachutist officers were drawn from the ranks and their training course usually lasted two months longer than that of the ordinary parachutist. The top-ranking parachutist officers were trained at special schools.

In 1939 Soviet airborne forces underwent their baptism of fire in the rôle of infantry. In the clashes between Soviet and Japanese ground forces along the Manchurian frontier, during the summer of that year, the 212th Soviet Airborne Brigade, stationed at that time in the Far East, was engaged in a series of bitterly contested local actions against the Japanese. This initial use of specially trained Soviet airborne forces in the rôle of infantry was a pointer to the general employment of parachute and airborne forces in the coming World War. For instance, while German airborne operations to capture Crete and the Dutch airfield at Waalhaven remain in the memory, in fact the bulk of General Student's or General Ramcke's airborne forces was more frequently engaged in stern defensive ground battles in North Africa, Northern France or on the Eastern Front against the Red Army.

But, when the war against Finland broke out in November 1939, Soviet parachute troops were used in the rôle for which they were intended, though in this campaign too airborne and parachute troops of the Leningrad Military District were engaged as part of the Soviet ground forces in the early fighting in Eastern Finland. However a force of probably not more than 2,000 parachutists was dropped ahead of the ground troops in order to seize objectives in front and behind the Mannerheim Line, and a smaller force was dropped near Petsamo in the north with the object of seizing Finnish-held strongpoints. The absence of wireless beacons or other navigational facilities and the poor standard of blind flying of the TB 3 air transport units, a crucial defect in the difficult flying conditions of a Finnish winter, the failure to reach the drop-zones accurately and to combine in combat after the drop—all these elements resulted in discouraging results. By December 1939 the Soviet High Command had decided to stop the parachutist operations and to deploy the troops again as part of the ground infantry forces. The loyalty of the Finnish population, the clever camouflaging and accurate shooting of Finnish infantrymen were other factors which prevented Soviet successes and discouraged them from using their parachutists for sabotage or guerrilla work in the Finnish campaign.

In the year before Hitler launched operation Barbarossa against the Soviet Union there was some fundamental rethinking going on among the top echelons of the Soviet Armed Forces. The lessons of the Spanish Civil War, the Finnish campaign and the Wehrmacht's phenomenal successes in France and Belgium in the summer of 1940, had been the subject of close study and much discussion. The inevitable conclusion was that modern war, as it then was, called for large armoured and airborne forces because it was so mobile. And so in the months before June 22nd, 1941, just as new tank corps were activated in the Red Army of which only a minority were ready for combat in June 1941, so the airborne brigades were expanded into airborne corps, but were inadequately trained, under strength and lacking the transport aircraft to use more than a small proportion of their existing strength. The new 3rd Airborne Corps had its headquarters in Kiev, the 4th Airborne Corps near Minsk and the 5th Airborne Corps in Leningrad. In addition a further airborne corps was created in the Far Eastern Provinces. Each corps was intended to have three parachute brigades. But in fact they had only one fully equipped brigade. They were subordinated to a Soviet Air Force which could provide few transport aircraft for their operations. And when the Wehrmacht achieved its striking successes in the summer of 1941, nearly all available transport aircraft were used to help evacuate Soviet industry to the Urals and beyond.

It is therefore not surprising that the Soviet airborne forces were engaged mainly in ground battles in the first year of the Germano-Soviet War. The new airborne corps were virtually deprived of transport aircraft because of the collapse of the Soviet air transport system in the first months of the campaign in Western Russia. Their units were thrown into the ground battles and inextricably mixed up with infantry divisions retreating towards Moscow, Leningrad and Kiev. One of the airborne brigades, the 214th, distinguished itself by breaking through a dangerous pincer movement near Smolensk and forming the nucleus of one of the new partisan groups formed following Stalin's rallying speech of July 7th, 1941. The use of Soviet parachutists to strengthen the partisan movement behind the German lines was to become a major facet of Soviet military policy in the Germano-Soviet War. But in the summer of 1941 the partisan movement was in its infancy and no doubt the surviving elements of the 214th Air Brigade were glad to get into civilian clothes, merge with the local population and await further orders from Moscow. Nevertheless despite the difficulty of retreat and acute shortage of transport aircraft, a small detachment from the much-battered 3rd Airborne Corps which had been fighting in the ground battles in the Ukraine in the summer of 1941 were dropped and linked up with the Soviet forces in besieged Odessa. Although only a few hundred reached the Soviet positions they were instrumental in strengthening what was a weak sector of the Soviet front at Odessa. The only other reported parachutist operations of 1941 consisted of a drop in about battalion strength from bases in the North Caucasus on the Crimean Front behind the German lines on the Arbat peninsula. This was in support of a Soviet counter-offensive in which their forces reoccupied the towns of Kerch and Feodosia. But in 1941 the bulk of airborne and parachutist formations had fought bitter defensive battles in the rôle of infantry and had suffered heavy losses, in particular the 4th Airborne Corps which had battled on the retreat all the way from Minsk to Moscow from July to December much as German airborne troops were to do in North Africa about a year later.

Nevertheless the Soviet General Staff decided that not only must the 1941 losses be replaced but new formations must be created. Even before the end of 1941 the decision to activate ten corps of airborne and parachutists troops was being implemented. The Caucasus and the Volga became the chief recruiting and training areas for the new corps which each numbered between 8,000 and 10,000 men, i.e. the target was up to 100,000 airborne specialist troops. As soon as the first corps was ready for operations it was despatched to the Moscow area as a support force for the coming Red Army winter offensive on the front west of Moscow. It was in this period that Soviet Airborne Forces ceased to be under the command of the Soviet Air Force but were made an independent arm as they are to-day, responsible directly to Stalin or rather to his Commissariat of Defence (to-day they are responsible direct to Marshal Malinovsky, Minister of Defence).

Once again it was the crack 4th Airborne Corps which was thrown into battle on the Moscow front behind the German defensive positions near Vyazma in January 1942. Soviet military sources say that this was the most extensive airborne operation of the Soviet-German War and that the entire airborne corps of some 8,000 men was engaged, with later reinforcements from units of the 3rd and 10th Airborne Corps, so that in all a force of 10,000 parachutists was dropped. German sources report the operation differently. They say that only one brigade of Soviet parachutists was dropped in a succession of night air lifts, that the drops were scattered and badly organised, that some of the men dropped without parachutes but that the airborne intruders continued to harass German positions for many weeks after the drop. In view of the limited numbers of twin-engined transport aircraft available, the Soviet parachutists were probably dropped south-east of Vyazma in waves of between 500 and 1,000 per sortie and since they suffered heavy losses and were reported to be dropped over a period of about a fortnight, their fighting strength probably did not exceed 5,000 at any one period. However, in sub-zero conditions with 40 degrees of frost (Centigrade) and ably led by Colonel Kazankin, they linked up with General Belov's raiding columns of Soviet cavalry, tank units and ski troops, as well as dropping inside the German lines and harassing the Wehrmacht headquarters at Vyazma in a series of night sorties. They fought on in the winter snows right up to the spring time in April and found reinforcements from paratroopers of the 4th Airborne Corps who had been fighting with partisans on this front almost since the beginning of the German-Soviet War. For those who think that there is a current Soviet plan to make an airborne assault on Alaska or Iceland, the operations carried out near Vyazma in the winter of 1941-42 confirm that post-war Soviet airborne divisions will not be deterred by near-Arctic climatic conditions.

Following the Vyazma airborne operations, an airborne assault was made in March 1942 on the German air base near Demjansk and the nearby German Army Headquarters. This operation was an important one and a brigade of Soviet parachute troops was used in an attempt to seize this major supply base which had been fed by German Junkers 52 transport aircraft for most of the previous winter. Although initial surprise was achieved, according to German sources, the coup failed, though it was a month before the parachutists were mopped up—a month of stiff fighting, and even then many escaped to join a partisan group and to harass German communications for the next year or so.

Despite the promise of the Vyazma operations German successes in the summer campaigns of 1942 again directed the bulk of Soviet airborne forces to the defensive ground battles of that year. At one time it had looked as though 1942 was to be a vintage year for the employment of Soviet parachutists and airborne troops, especially when the long-range combined bomber and transport air command (the ADD) was reformed in the spring to provide a separate air army of these machines for strategic use. But because of the threat of the German summer offensive and the need for an élite force of ground troops to oppose them, the Soviet General staff decided to withdraw the bulk of the ten airborne corps from their specialist work and to re-deploy them to stiffen existing infantry and tank formations. They were then rushed into the hard-fought battles on the Don and Stalingrad fronts. However, before the end of 1942, three useful airborne operations were carried out on various southern fronts in the German-Soviet War. Many thousands of crack infantry were air-ferried to Stalingrad by PS 84 and TB 3 aircraft to relieve the pressure from the Sixth German Army in October 1942; an airfield was seized against weak German opposition near Maikop, in the Caucasus in the same month; and a small force of parachutists was dropped near the Black Sea port of Novorossiisk. Unfortunately no details are available on the tactics employed in these last two operations, nor of the precise strength of the Soviet airborne forces engaged.

It was clear by the spring of 1943 that the Soviet High Command were unwilling to leave a large body of picked infantrymen lying fallow for long periods until the combat situation permitted their employment as airborne troops. No doubt the uncertain supply of transport aircraft for large-scale operations was a further factor which persuaded the Kremlin during this period to convert the ten so-called airborne corps to ten "Guards Airborne Rifle Divisions". However, the name airborne stuck to these units up to the summer of 1945, as no doubt the distinctive title added to their prestige and morale. In the last two years of the Second World War they fought prominently in many of the land battles, including the capture of Budapest and Vienna and the crossing of the Oder.

Nevertheless specialist airborne and parachute troops were still a major tactical arm in the strategy of the Soviet High Command. In the spring of 1943, as the main body of the airborne corps was being converted to Guards Airborne Rifle Divisions, the Commissariat for Defence was charged to create Guards Airborne and Parachute Brigades each with a strength of about 3,000. While the Soviet intention was to form fifteen to twenty of these brigades by the end of 1943, it is believed that not more than ten of them were fully trained and equipped at any one time and the air lift resources available from the ADD Long-range Air Command and the Aeroflot Civil Air Lines did not permit the simultaneous employment of more than about 5,000 parachutists or airborne troops in one operation. For this some 200 serviceable transport machines, mainly PS 84s, were needed, each carrying about twenty-five men and their equipment. This was about the maximum number that the Soviet air resources could provide, that is less than half the number of transport planes employed in major German airborne and parachutist operations of the Second World War, and less than one-half of the resources deployed by U.S. air transport units in the Sicily airborne landings in the summer of 1943 or the airborne landings in Holland in September 1944.

The new Soviet Guards Airborne Brigades each consisted of four parachute battalions of some 750 men, each battalion having special engineer, anti-tank, light artillery and machine-gun units. The first major operation they were asked to carry out in 1943 was in fact the last occasion that the Soviet High Command decided to use parachutists in a large-scale operation in the Second World War. In September, some 5,000 élite parachutists chosen from the 1st, 3rd and 5th Guards Parachute Brigades and commanded by Colonel Sidorchuk were dropped across the River Dnieper near Kanev, south of Kiev. But the operation was not very successful and there was little to show the Soviet High Command how crucial parachutists could be as a major aid to the establishment of bridgeheads across rivers. The other river crossings of the Dnieper further south, the crossings of the Dvina and the Oder were achieved as smoothly as the Kanev crossing without the aid of Soviet parachutists. It is incidentally a a matter of military speculation whether the British crossing of the Rhine in the spring of 1945 would not have been just as expeditious without the support of an airborne division in the operation. At all events, despite the many opportunities for dropping airborne forces ahead of the assaulting Soviet ground troops in the last year to eighteen months of the German-Soviet War, for example in the 1944 assault on the Crimea, or in the recovery of Polotsk, Vitebsk or Orcha, despite the availability of large numbers of parachutists and reasonable numbers of transport aircraft, there were no more air drops in divisional or brigade strength. It would seem that such blitzkrieg thrusts as those of Malinovsky and Tolbukhin across the rivers Dniester and Prut in 1944 made airborne support in bridgehead operations entirely superfluous. The last recorded operations were small-scale drops in August 1945 in the brief Soviet-Japanese War in Manchuria. It would seem that the Soviet airborne operational talents were best suited to the less ambitious tactical rôle of dropping small groups of saboteurs who co-operated with partisan forces on the lines of the U.S. "special" airborne services and British S.O.E. operations. The great difference was of course the greater scale of the Soviet airborne operations in this field. For instance in 1943 some twenty-five German divisions were engaged in fighting the Soviet partisans apart from SS and SD German police formations, that is some ten per cent of all ground forces engaged against the USSR. In a single month in that year the Wehrmacht reported over 1,000 railway demolitions by Soviet partisans and over 4,000 attempts to sabotage trains and rail trucks were discovered and prevented. Many thousands of German-driven railway engines were damaged by Soviet partisan activity in the 1942-43 period, and some tens of thousands of rail coaches. In addition the Soviet partisans operating in the rear of the German front were able to provide from time to time military intelligence of great importance as well as prevent the German occupiers of Soviet territory from exploiting fully the economic and agricultural wealth of the territories they held.

It would be wrong to pretend that the bulk of the partisan forces was parachuted into their zones of operation or were rump remnants of airborne units which had been engaged in the ground battles. Many of them were Soviet civilians who wanted to avoid German punitive measures in the USSR and the threat of forced labour in Germany. But their arms, supplies and leaders came by Soviet transport aircraft and so made the partisan fighters an important part of Soviet war plans from 1941 to 1944.

Looking back on Soviet policy in their use of parachute and airborne troops in the Second World War, one finds it severe and practical, as are so many aspects of Soviet military policy. It was ready to ignore the full potential of a new air arm in which it had pioneered when it seemed likely to be abortive or too expensive to retain in idleness for too long. This showed that the Soviet High Command was realist, wise and flexible, ready to depart from the orthodox use of parachute troops when the homeland was threatened or when common sense demanded. The difficult question now is to decide what uses the Soviet High Command are likely to make of airborne and parachute forces in the unlikely event of a major war in the atomic, rocket and guided-missile age. Despite the lack of outstanding operational successes and outstanding commanders of airborne forces in the Second World War, Soviet faith in their importance in modern war has not been shaken. The general view is that they have some ten full-scale divisions of parachute and airborne troops with a strength of some 100,000 men. In addition they have fostered and trained much smaller forces, usually about a division, of airborne and parachute formations in the armed forces of all their European Communist allies as well as in Communist China. Soviet parachute divisions form an independent arm working directly under the Ministry of Defence, where the senior officer is believed to hold the rank of marshal. The airborne forces have a separate headquarters and administration for aircraft and for equipment, supplies and political administration. Their aircraft equipment is more modern and extensive than at any other period in Soviet military history. The transport planes and helicopters are described in detail in the special chapter dealing with these machines. All that is noted here is the great progress made in the 1950s in developing powerful helicopters such as the YAK 24 and the MIL 6 which can carry large numbers of troops with equipment such as anti-tank artillery and machine guns, trucks and motor cycles, thus ensuring a high mobility in supporting ground troops. Since 1953, the airborne troops have been carrying out annual manœuvres, using at first between 25 and 50 large helicopters and later over 100. It is tentatively estimated that in the early 1960s the airborne forces will have available a force of perhaps 300-500 large helicopters, a mixed force of YAK 24 and MIL 6 machines which will enable it to lift 10-15,000 men and their equipment in one tactical operation. The latest helicopters are being fitted with automatic pilot devices and the pilots are having training in blind flying so that day and night and good and bad weather operations are feasible. To judge from details of the training and manœuvres available it would seem that some units of Soviet airborne troops will be used ahead of the front line to seize targets or vantage points; and published statements in Soviet military journals indicate that the Soviet High Command attaches great importance to the employment of airborne troops as reinforcements in ground warfare where atomic weapons are used. Clearly in such military operations road communications may be uncertain and many local combat areas difficult to reinforce or supply by any means other than by air. Thus the airborne forces are likely to play a more crucial rôle in ground battles in the future than in the past.

The ten or so Soviet divisions of airborne troops each break down into a regiment of parachutists, one of air landing troops and finally a regiment of special sabotage troops who are experts in guerrilla combat. They are all élite troops of the high category that would make them non-commissioned officers in ordinary Soviet ground formations. They are fitter and more intelligent than ordinary soldiers. The standards of parachute-jumping, weapon-handling and map-reading are reported to be high and cross-country running, unarmed combat and ju-jitsu are part of the normal training programme. They receive of course better rations and pay than ordinary Soviet ground troops.

In view of the special organisation of sabotage and guerrilla fighting regiments, one can reasonably assume that the Soviet High Command plans to use parachutists on these operations on an even greater scale than in the Second World War. In Western Europe, for instance, they might be used to blow up rocket missile batteries in combination with local Communists who might be able to pinpoint the target for the Soviet airborne saboteurs. They may also be used as in World War II to harass and damage air bases as they did in the Germano-Soviet War with such success that the Luftwaffe had to employ special airfield guards as an extra protection. If the ground fighting were in the Middle East, they might be used in similar airfield attacks, in operations against oil pipe lines and in arming and training local Communist elements, e.g. in Irak.

But the greatest changes in planning potential derive from the extraordinary developments in the production of Soviet long-range jet transport aircraft in the 1950s. We hear of the Tu 104 twin-jet, the Tu 110 four-jet and the four jet-prop Ilyushin 18, Antonov 10 and Tupolev 114 as the new passenger planes of Aeroflot. In war they are likely to be placed largely at the disposal of Marshal Skripko who commands military transport aviation. This means that if necessary a two- or three-division air lift to seize an air base in Iceland, or to bolster Communist forces in the Middle East may well be feasible in the early 1960s. And as Soviet long-range jet transport machines increase in performance, numbers and range, as they undoubtedly will in the next five or ten years, so their potential to operate against Britain, U.S. bases in Spain and even Greenland and Alaska, will gradually become so great that it can be considered a serious aspect of Soviet military planning. The progress in jet transport aircraft in the late 1950s has done much to make the entire military forces of the USSR more mobile, effective and flexible in the atomic age.

How important this can be in modern combat is emphasized in General James Gavin's recently published book *War and Peace in the Space Age* (Hutchinson, London, 1959). Gavin writes: "Between now and 1965 the form of mobility of most significance to us will be air mobility . . . air transport is not only tactically desirable but administratively essential . . . fire brigades must be flown, for in dealing with limited wars the amount of force required is a function of time." The military leaders in the Kremlin are no doubt aware of the soundness of these tenets and will strengthen their airborne arm even further in the coming decade.

CHAPTER 7

STRATEGIC AIR DEFENCE

IT is probably true to say that, relatively speaking, strategic air defence has always been the weakest part of Soviet air power, that it is so to-day and is likely to remain so in the future. Until the end of the Second World War, the radical weaknesses in Soviet air defence against long-range air attack, e.g. the absence of an early warning radar system, were not serious factors in their overall military potential. Indeed it was not until the end of the 1939-45 War that a strategic long-range air threat of any magnitude developed. Although Germany had built and designed a number of promising four-engined heavy bombers such as the Dornier 19, Junkers 90 and Focke-Wulf 200 prior to World War II, none of them went into large-scale production, and so the strategic air threat to the USSR from the Luftwaffe was never very great. Once the war began, Germany's chief bomber effort was in tactical support of her ground forces or in attacks on British and later American shipping in the Atlantic, the North Sea and the Mediterranean. And after Pearl Harbor, the Japanese preoccupation with the campaigns in the Pacific and South-East Asia virtually eliminated any threat of long-range air attack on the USSR. In any event Japan was not strong in her strategic bomber arm.

The Soviet High Command had early evidence of the strict limitations of the scale of German long-range air attack on Soviet towns and industries during the Second World War. When in the summer of 1941 the Soviet had to move its aircraft, tank and other industries from White Russia and the Ukraine to the Urals and beyond, they did so without any hindering action from German bomber forces. And when Moscow came under German bomber attack in July and August 1941 the Luftwaffe actually had to transfer special Heinkel pathfinder bomber units from Western France to Poland and to pull out other bomber units from the tactical air battles in the Smolensk and Kiev fronts to mount the air assault on the Soviet capital. Even so the major raids by the Luftwaffe on Moscow in this period never exceeded 200 sorties and rarely 100, and within a week or so were petering out to raids of about 50 night bombers, which as Soviet fighter and anti-aircraft opposition increased and improved became almost entirely ineffective. For the rest of the Germano-Soviet War there was no other instance of any sustained long-range attack on Soviet cities by German bomber forces. In the summer of 1943, the Germans did carry out a series of long-range attacks on Soviet factories behind the front line at Gorki, Jaroslav, Saratov and Astrakhan. These attacks went on for about a fortnight in the second half of June, but early in July they ceased abruptly as the Heinkel 111 and Junkers 88 bombers engaged were urgently switched to the tactical air battles in support of ground operations on the Kursk, Orel and other fronts. And as far as the rest of the Germano-Soviet air war was concerned, the only Soviet air defence commitments were of a local tactical nature, such as meeting the concentrated German air attacks on Leningrad in the summer and autumn of 1941, and those on Odessa and Stalingrad in 1942.

In general it is fair to say that the Soviet Air Force reached the end of the Second World War with its strategic air defences virtually untested. Nevertheless one can legitimately assess some aspects of the strength and weaknesses of Soviet potential in strategic air defence during the Germano-Soviet War. Perhaps the most striking weakness was the total absence even at the end of the war of any early warning radar system, without which modern air defence is likely to be virtually impotent. What a handicap this might have been, must have been made clear to the Soviet High Command in the early days of the Barbarossa campaign when the absence of reliable local early warning systems permitted Luftwaffe bombers to make unheralded and devastating attacks on Soviet-held airfields and other targets before Soviet fighters and anti-aircraft guns could be brought to bear. The Soviet early warning system used in World War II was based almost entirely on visual observation and to some extent on sound detectors, with all the uncertainties of these methods especially in cloudy weather or at night. Some of the uncertainty was graphically demonstrated in the summer of 1944 when U.S. Flying Fortresses were temporarily based at an airfield near Poltawa and suffered heavy losses and damage on the ground during a night attack by some fifty German Heinkel planes, partly because the Soviet early warning system gave little indication of the approach of the Luftwaffe forces. Eyewitness accounts by British air personnel who spent the winter of 1941-42 on the Murmansk front training the Soviet to assemble and fly British Hurricane fighters, and the well-documented evidence of General de Gaulle's Normandie-Niemen squadrons, which flew Soviet YAK fighters for long periods of the Germano-Soviet War from 1943 to 1945—all the evidence from these sources leaves one in no doubt that the Soviet early warning system was haphazard and often primitive, but fortunately rarely submitted to a full-scale test.

The strongest part of Soviet air defence in the Second World War was, relatively speaking, the artillery arm. Stalin called his artillery the "God of War" and his chief Marshal of Artillery, Voronov, was one of his closest military advisers, almost a friend. Stalin described his artillery as "the main striking force of the Red Army", and this provides the key to its main use. Like Soviet fighter aircraft the artillery brigades, regiments, battalions and batteries were used mainly within the framework of the tactical ground battles, as anti-tank artillery or for mass fire in support of local armoured and infantry offensives. It is therefore not surprising that Soviet anti-aircraft gun batteries were in quality and experience well behind those of Germany and the Western Allies, though on the other hand, many military critics have stated that Soviet anti-tank guns were better than those of the German. While Goering exercised great control over the use and deployment of Germany's dual purpose medium and heavy artillery and used a high proportion of it for anti-aircraft defence, especially from 1942 to 1945, Voronov was only prepared to release for local air defence such Soviet artillery and heavy machine guns as he thought could be spared from the ground battles on the Eastern Front. And because Soviet artillery was in the main intended as one of the chief supporting arms to the Red Army, the development of rockets and mortars for Soviet air defence was hampered and restricted in World War II, although considerable use of these weapons was made by Soviet air and ground units in both the defensive and offensive tactical battles of the German-Soviet War. In the combat organisation of Soviet artillery the anti-aircraft units could in case of need be assigned ground support rôles as anti-tank or field artillery if the need arose. This flexible dualism was true of German anti-aircraft flak batteries too, but in the Soviet case the diversion from air defence to tactical ground support of the army occurred much more frequently.

Nevertheless in a number of witnessed instances, Soviet anti-aircraft fire in World War II showed that it could be an effective local deterrent to air attack. British and U.S. observers in Moscow, who were subjected to Luftwaffe air attack by Heinkel and Junkers bombers in the summer of 1941 have testified to the intensity and concentration of Soviet anti-aircraft fire in defence of the capital during that period. An examination of German air documents reporting Luflotte 2 operations on the central Eastern Front in July and August 1941 indicate clearly that the strength and accuracy of Soviet anti-aircraft fire were occasionally both a surprise and a deterrent to the raiding German bomber crews. And further testimony to Soviet anti-aircraft fire can be found in German reports on air attacks on Leningrad and Stalingrad. Then there is both the evidence and the tribute of the R.A.F. Hurricane wing which occupied bases in the Murmansk area in the winter of 1941-42. Nevertheless the Soviet anti-aircraft defences were never subjected to persistent long-range air attack in the Second World War, and their potential compared with, say, British or German anti-aircraft gun defences cannot be fundamentally assessed.

The same is of course true of Soviet fighter defences. There was never any serious likelihood of Soviet need to build up a strong force of strategic defence fighters as Germany and Japan had to do; and so only about one fighter division was permanently assigned to the continental air defence of the USSR in both the pre-war and war period. It had a strength of perhaps 250-350 fighters. Nevertheless the Soviet authorities realised that the uncertain political situation in Europe in the 1930s might well lead to air combat with or against almost any combination of the great powers; and so they began to specialise in the development of target defence interceptor fighters. In 1934 TSAGI appreciated as early as any other contemporary technical authority the importance of rocket and jet engines to boost the speed and performance of fighters engaged on bomber interception operations. Tupolev mounted a series of solid fuel rockets under the wings of an experimental fighter, the I 4. As the Second World War approached, three rocket fighters were designed for target defence and experiments, and tests were carried out in 1939 and 1940 with Merkulov ram-jets fitted under the wings of the then current operational fighters, such as the I 15 biplane designed by Polikarpov and a YAK monoplane fighter. However, because of a loss

A ROW OF FREIGHT ROCKETS READY FOR LAUNCHING.
FROM THE FILM "ROAD TO THE STARS" 1957

JET EQUIPMENT ON RED SQUARE, MOSCOW, 1957

RUSSIAN ROCKETRY

THE MIL-6, HOLDER OF THE WORLD HELICOPTER SPEED RECORD IN 1959

AT A S.A.F. AIR DISPLAY, TUSHINO AIRFIELD, MOSCOW, 1955

SOVIET HELICOPTERS

of manœuvrability and general combat qualities in the Soviet fighters when the ram-jets were fitted, and because the speed boost obtained was only of the order of ten to twenty per cent, the experimental use of the Merkulov ram-jets was soon discontinued. In the same period, that is a year or so before Hitler launched his attack on the USSR, TSAGI also sponsored a useful rocket engine designed by Dushkin giving a modest thrust of some 700-850 lb. The Dushkin rocket power plant was actually installed in three Soviet prototype target defence interception aircraft and after a series of test flights, a high-powered Soviet defence committee selected one of them, the BI 2 designed jointly by Berezniak and Isaev, to be put into small-scale production from the summer of 1941. Meantime Dushkin's rocket motor had been boosted to over 1,000 lb. thrust and the endurance had been raised from about 10 minutes to 15 minutes. The new Soviet rocket fighter had a nose-mounted armament of two 20 mm. machine guns. Its development test flying in 1942 was successful but its endurance was probably too limited for operational purposes. And so by 1943 the plan to develop a Soviet rocket fighter interceptor was abandoned. The subsequent history of rocket fighter aircraft in other air forces suggests that the Soviet decision at the time was correct. It is true that Germany went on to develop an Me 163 rocket interceptor fighter which had limited combat experience against British and American bombers in 1944 and 1945, but its operational record was poor. It was difficult to fly and cost the lives of some of Germany's best fighter pilots without producing any important results against the U.S. bombers and fighters to which it was opposed in 1945. Indeed the development of fighters in the jet and rocket decade that followed World War II pointed to the need for both jet and rocket motors in one aircraft in the age of supersonic flying, simply because an engine that breathes air can scarcely be effective at heights where the air is thin; and it is then that the rocket engine takes over from the jet.

When the dust of Germany's defeat began to settle in the summer and autumn of 1945, it became clear to the Soviet General Staff that a radically new approach to air defence was needed in the new jet-atomic age, in which a force of long-range bombers could destroy Soviet military potential in a few days or even in a few hours. Stalin and his senior military colleagues were much impressed by the great damage caused by British and American heavy bombers to German and Japanese cities and industry. The lesson for the USSR was clearly to lay greater stress on strategic air attack and defence. And so a long-term plan was devised to strengthen the fighter and gun defences of the USSR against atomic air attack. The plan was bound to be long-term, for it would clearly need several years at least to produce an efficient early warning radar system, which is the indispensable prerequisite of strategic air defence by fighters, guns or missiles. To manufacture the radar equipment, to select and train the personnel for an early warning system covering one-sixth of the globe was a new and gigantic undertaking. In this and other aspects of the new post-war Soviet plan for strategic air defence the Kremlin's debt to German technicians, equipment and blue-prints has been immeasurable. Among the wide range of German electronic assets acquired by the Soviet Air Force in the 1945-46 period were many usable samples of German early warning Freya and Wurzburg apparatus which had been in production in East German and Polish factories and which could not be evacuated or destroyed in the final retreats and defeats of the spring and summer of 1945. To add to this equipment windfall, there were also many hundreds of trained Luftwaffe radar personnel who had manned the German Air Force early warning system set up in Roumania, Hungary, Poland and East Prussia in the last year or two of the war. These specialist signals personnel were sifted from the masses of German prisoners of war and sent to train members of the Soviet Air Signals units at Leningrad, Kiev, Riga, Moscow, Tashkent and elsewhere in the arts of servicing and operating the German-designed early warning radar equipment. Add to this windfall some further hundreds of German radar engineers with production and repair experience at the German electronic firms of Siemens, Askania and Telefunken, and one begins to see why the first Soviet skeleton radar early warning system covering the approaches to the East Baltic and to Eastern and Central Europe was set up as early as the end of 1946, the year incidentally in which German electronic and radar technicians, who had moved from West German factories eastwards between 1943 and 1945, now began a further trek in the same direction which took some of them to Soviet factories east of Moscow and indeed as far east as Vladivostok. Air Marshal Kliment Vershinin, at this time Commander-in-Chief of the Soviet Air Force, pushed on rapidly and urgently with the work of developing a continent-wide early warning system for the USSR. By the outbreak of the Korean War in 1950, the radar stations had expanded to the Far Eastern Maritime Provinces, including the vital Magadan region opposite Alaska, as well as to the coastal areas of the Black Sea from the Crimea down to Batum.

All this was but the beginning of a comprehensive strategic air defence system which became more and more essential as the stockpile of U.S. atomic bombs grew. The Soviet General Staff realised that it would take years—perhaps more than a decade—to produce enough radar equipment, both early warning ground equipment and airborne equipment for ground-controlled interception by aircraft, quite apart from manning the equipment and training the personnel before the USSR would have an efficient early warning system in depth and covering all parts of the Soviet Union.

One of the most difficult things to assess both for the Kremlin and for the Western powers is the potential efficiency of the Soviet early warning system, whether it is operating in support of the present-day supersonic jet fighters and guided missiles or for the anti-missile missiles of the future. As far as can be judged by the practical evidence of the last ten years or so, Soviet radar equipment is manned in general by reasonably skilled signals personnel. The Korean War provided the first post-war large-scale test for Soviet-manufactured radar equipment manned by Soviet signals personnel. In general they came out of some three years of campaign experience with an enhanced reputation. The Soviet jet MIG 15s, supplied in large numbers to the Chinese Communist Air Force, owed the tactical height advantage which they frequently enjoyed in the air combats near the Yalu river to the adequate and consistent early warning of the approach of United Nations bombers and fighters supplied by Soviet-manned early warning radar units. These often enabled the Chinese-flown MIG 15s to be airborne in time to gain a height advantage and to dive out of the sun on their U.S. and British air opponents. It was the higher training standards and superior pilot qualities of the United Nations aircrews which were the chief reasons for the relative lack of success of the intercepting Chinese-flown MIG 15s.

However if the Korean War showed that the Soviet Air Force and aircraft components industry had satisfactorily adapted and used the German ground early warning radar equipment, it would seem that the adaptation and large-scale manufacture of airborne radar, such as the German Naxos, Lichtenstein and SN 2 equipment, was taking the Soviet longer to achieve, to install in their fighters and to operate. There was no sign that this equipment was used on any scale in the Korean War when it would have been invaluable on night-fighter operations against United Nations B 29 bombers, nor did the MIG 15s have a radar gunsight, a disadvantage which partly explains the combat successes of the U.S.-flown Sabre F 86 jet fighters and the rather poor shooting qualities of the Chinese-flown MIGS. And since the Korean War became a testing ground for Soviet weapons and equipment of strategic air defence, it is noteworthy that none of the guided ground-to-air or air-to-air missiles developed by the Luftwaffe in 1944-45 were ready to be tried out in the 1950-53 war in Korea. No doubt such German weapons as the Wasserfall, Schmetterling and Rheintochter ground-to-air missiles needed a longer period to test, to modify to Soviet needs and then put into large-scale production and finally to large-scale use. Evidence obtained subsequent to the Korean War suggests that the first operational batteries of Soviet ground-to-air guided missiles did not reach the PVO Strategic Air Defence Command until 1955 when they were allocated to the defence of the Moscow region.

However it was the development of jet fighters which marked the main progress in the Soviet weapons of strategic air defence in the decade that followed World War II, just as the development of guided missiles, in particular the anti-missile missile is the main air defence feature of the current decade. During the Second World War the Soviet High Command created in 1944 a special Fighter Command in the PVO organisation which is referred to in Soviet air documents as IA-PVO (I stands for *istrebitel* and means fighter, and A for *aviatsia*—aviation). Experimental rocket, ram-jet and turbo-jet engine developments were continued by such engine designers as Dushkin and Lyulka, and the latter is to-day one of the most successful of Soviet jet-engine designers. Once again the German legacy, and later the sale of British Rolls-Royce jet engines to the USSR, helped the Soviet to close the technical gap between the USSR and the West in the field of jet-fighter planes in the early post-war period. The capture of large numbers of German Junkers Jumo 004 and BMW 003 co-axial jet engines speeded up the Soviet post-war plans to develop jet fighters for strategic air defence. The first two Soviet post-war jet fighters, the MIG 9 and the YAK 15, were test-flown by April 1946. According to Soviet sources, the twin-engined MIG 9, designed jointly by Mikoyan and Gurevich and powered by Soviet RD 20 jet engines, virtually copies of the German BMW 003, reached a top speed of over 560 m.p.h. with a service ceiling of over 42,000 feet. These performance figures, if accurate, would put the new Soviet jet into roughly the same class as the contemporary British Vampire and the U.S. F 80 Shooting Star. However reports suggest that the RD 20 engines were not reliable and there were other vital factors such as pilot training, radar gunsights, armament and radio equipment which need to be taken into account before valid overall comparison can be made between Soviet and Western jet fighters as gun platforms for airborne interception.

But there was continuing technical progress in the development of jet fighters and in the selection of élite pilots, in the late 1940s as well as numerical strengthening of the IA-PVO, the fighter arm of the Soviet strategic air defence. Alexander Yakolev's YAK 15 plane, essentially an adaptation of his YAK 9 piston-engined fighter re-powered by a Soviet edition of the German Junkers Jumo 004 axial jet engine, had a short life. First flown in the spring of 1946 it was soon to be replaced by the YAK 17, with an improved turbo-jet RD 10 engine designed by the Lyulka bureau. This was in the 1947-48 period before the Korean War. It was during this period that the Soviet began to develop modern swept-wing jet fighters for the coming supersonic age of flight. Lavochkin produced the LA 160 with a 35 deg. sweepback which attained in flight trials the equivalent of Mach 0·92 in level flight, but this aircraft never went into large-scale production. The fact is that the USSR was not yet making turbo-jet engines of sufficient power to enable her to keep pace with Western jet fighters or to produce engines powerful enough for modern jet bombers. A jet power unit giving 4,000-5,000 lb. static thrust was called for and neither the Lyulka nor the Mikulin engine design bureaux working on this problem were able to produce it. And so the USSR turned to Great Britain and bought from her 25 Rolls-Royce Nene and 30 Rolls-Royce Derwent turbo-jets early in 1947. The two British jet engines were an important contribution to the modernisation of the Soviet jet fighter arm in the late 1940s and perhaps early 1950s. But the overall achievements of the Lyulka, Mikulin and Kutznetov design bureaux in this field in the last decade suggest that the Rolls-Royce trading venture did no more than advance the appearance of the MIG 15 and perhaps the MIG 17 and the YAK 23 by about a year.

The two British engines, the Nene (the Soviet edition was called the RD 45) and the Derwent (RD 500) were by the end of 1948 beginning to power some of the more advanced Soviet jet fighters. The YAK 23 was powered by an RD 500 which gave it a top speed of 568 m.p.h. and a service ceiling of over 48,000 feet. It was to be much less successful than the MIG 15 which was probably the first Soviet jet aircraft to become widely known in the West. First flown in December 1947, it was the first jet fighter to be produced in real quantity in the USSR, that is in thousands per annum. This occurred at the end of 1948 and by the outbreak of the Korean War in June 1950 the Soviet strategic air defence fighter arm of over 2,000 jet fighters was extensively equipped with this new fighter plane.

Moreover in the five years following the end of the German War, the Soviet artillery arm of the P.V.O. had been substantially strengthened and improved by the addition of hundreds of radar-controlled heavy anti-aircraft guns of up to 120 mm calibre, which, like the greatly improved searchlight equipment were mainly guided by radar. Some progress had also been made in deploying a few radar-guided ground-to-air two-stage anti-aircraft missiles, based largely on the German Wasserfall design. And at the same time the early warning radar screen in Eastern and Central Europe, over the Baltic, the Arctic and the Far Eastern Maritime Provinces was being gradually extended and improved. There can be little doubt that if the Soviet strategic air defences had had to go into action during the Korean War, they would have given a fairly good account of themselves compared with the U.S. continental air defences. It was possibly the lack of a long-range jet bomber force and the lack of a big stock of atomic bombs which made the Kremlin hesitant to commit itself to a major conflict. The MIG 15, which before the end of the Korean War, equipped nearly all the fighter squadrons of the IA-PVO air regiments, compared favourably in many ways with its U.S. counterpart the F 86 Sabre or indeed with any jet fighter in squadron service anywhere in the world at that time. With a top speed of over 675 m.p.h. and a service ceiling of 50,000 feet, it was faster than the U.S. Sabre at heights of over 35,000 feet, could turn in a smaller circle in combat and had a faster climb rate. However, its instability at high speeds and the lack of a radar gunsight made the MIG 15 an inferior piece of shooting equipment. The lack of airborne radar and a radar gunsight was a major defect in Soviet strategic air defences at this time; but by 1953, when the Korean War ended, much had been done to remedy these deficiencies. The Soviet by then had test-flown its first jet all-weather fighter, the subsonic YAK 25. This 685 m.p.h. (Mach 0·95) aircraft began to equip the day and night defence air regiments of the PVO command in 1955 and to-day is still the backbone of the Soviet night-fighter regiments, though it is being gradually supplanted by the latest Soviet ground-to-air guided missiles. The YAK 25 has a radius of action of 750-1,000 miles, but this twin-jet fighter, powered by engines which now with afterburner give some 10,000 lb. thrust each, is likely to be replaced by a supersonic all-weather fighter, which as long as a strong U.S. Strategic Air Command with a manned bomber force exists, can provide more flexible and effective overall opposition than the current generation of Soviet gound-to-air guided missiles limited to a range of only some 25-50 miles. The YAK 25, dubbed by NATO the Flashlight, has a top speed of about 720-725 m.p.h. It may have already been partly replaced by a supersonic twin-jet akin to the NATO-dubbed Blowlamp, which was reported to be test-flown as far back as 1956.

Together with the development of long-range twin-engined all-weather fighters, the single-engined day-fighter equipment has been progressively improved since the early 1950s to the present point when the MIG 15 is almost entirely obsolescent and the day-fighter air regiments of the PVO are almost entirely equipment with supersonic jet fighters. The MIG 17 began to supersede the MIG 15 in 1953. It was fitted with a more powerful VK 1 Klimov turbo-jet engine, giving initially a thrust of some 6,000 lb, and later, with an afterburner fitted in the middle 1950s, a thrust of some 7,000 lb. without achieving supersonic speeds. It went into large-scale production in the middle 1950s and equipped not only the strategic fighter and tactical fighter air regiments of the Soviet Air Force, but also the naval air units of the Soviet Navy and the air squadrons of Poland, Czechoslovakia, the United Arab Republic, Communist China and Afghanistan. One version of it is an all-weather fighter with airborne radar and another carries extra tanks giving it a radius of action of some 500 miles. With a speed of Mach 0·97, that is about 645 m.p.h. at 50,000 feet and over 725 m.p.h. at low levels, and a service ceiling of about 58,000 feet, the MIG 17 enabled the Soviet Air Force to continue to keep pace in jet-fighter development with the West in the middle 1950s, and before the supersonic jet and rocket fighter and guided-missile period emerged in the late 1950s.

Certainly by 1955 the Soviet Air Force had all the major elements of a first-class strategic air defence organisation. In that year not only did the YAK 25 twin-engined subsonic all-weather jet fighter enter service with the air regiments of PVO, but also the first supersonic day fighter, the MIG 19, the Soviet counterpart of the U.S. F 100 Supersabre, began to equip the PVO fighter regiments. The new jet fighter was capable of some 925 m.p.h. in level flight, equivalent to a speed of about Mach 1·4. It was powered by two axial flow jet engines designed by the Lyulka bureau, giving thrusts of over 6,500 lb. each and nearly 9,000 lb. with afterburner. The early MIG 19s of the 1955-56 period were armed with two 23 mm. cannon in the wing roots, and one larger 37 mm. Nudelmann cannon firing forward from the port side of the fuselage. In addition eight unguided rockets, each of 50 mm. calibre, could be carried. It was also in 1955 that there was evidence of the installation of radar gunsights in Soviet jet fighters, as well as the experimental use of air-to-air missiles and an expansion in the number of ground-to-air guided missiles, batteries probably fitted with T 7 and T 8 type missiles, the former reportedly having an infra-red detection system as well as a radar-tracking target system linked by computor to missile-tracking radar. Finally the Soviet early warning radar system was now reasonably comprehensive and provided warning in depth through an elaborate three-tier line of stations ranging from the Arctic coasts in the north down to the Warsaw-Omsk-Novosibirsk and Irkutsk line, with a southern line covering the approaches to the USSR via the Balkans, the Black Sea, Persia and the Caspian. There was also the eastern line stretching south-westwards from the Bering Sea via the Maritime Provinces to the Afghan frontier.

But perhaps the most significant development was organisational. The PVO which had been a neglected Cinderella of the Soviet armed forces before and during World War II now became one of the leading military arms of the Soviet defence forces, when Marshal Biryuzov was appointed as its Commander-in-Chief, and a deputy to the Soviet Defence Ministers, at first to Marshal Zhukov and later to Marshal Malinovsky. Biryuzov had had limited operational experience in the Second World War when he was an artillery commander with the Far Eastern air and land armies. It was not until 1944 that he was transferred to the West to fight on the Belo-Russian front against the Wehrmacht. When the European war ended he was again given a Far Eastern artillery command and his gunners took part in the brief campaign in Manchuria during which they achieved successes against the Japanese 5th Kuantung Army. Holding the rank of Colonel-General, he was transferrred in 1945 to the artillery department of the Soviet General Staff and given the job of planning the expansion and modernisation of the post-war PVO which he now (1960) commands. A heavily built man of about 60, Biryuzov is a typical member of the Soviet military political aristocracy, energetic and devoted. During the Second World War one of his closest friends was Lieutenant-General Nikita Khrushchev, now the Soviet Prime Minister and virtual controller of the Communist Party machine in the USSR.

The important new status of Biryuzov's PVO anti-aircraft and defensive missile command is clear not only from the fact that the Marshal is a deputy Defence Minister, and so at least equal in status in the Soviet military hierarchy to any of the Commanders-in-Chief of the other air or ground arms; in addition Biryuzov has had overall influence and authority for his PVO command in all aspects of anti-aircraft defence, i.e. policy, administration and decision. Marshal Biryuzov controls and commands the fighter regiments, gun and missile batteries, signals units, in fact everything concerned with active air defence. For nearly forty years the Soviet Army had exercised control over defence matters regionally. All Soviet forces within the region were subordinate to the Army Commander. But since about 1955 the PVO has enjoyed the privilege of complete local control over its forces with its own independent headquarters. Some times the regional PVO command works side by side with the local army command, as in Baku and to some extent in Moscow. In other areas, for example north of the Arctic circle, there is no military, i.e. army, region as such; in the case of this vital northern first line of strategic air defence against long-range air attack across the North Pole, the PVO is closely linked with the DA long-range attack forces of the USSR. There are in all more than twenty regional PVO commands each headed by an army or air force general. They cover every area of the Soviet Union. In the Arctic North PVO headquarters and regional commands exist near Murmansk, in the Talmar peninsula and further east in Kolimsk and the Kamchatka peninsula. There are also PVO commands near Leningrad and Moscow, in the Crimea, at Baku, and Astrakhan, Omsk, Stalinabad, Komsomolsk and Vladivostok.

What developments have there been in the Soviet strategic air defence forces during the past few years and what is its potential in the future? Technically, numerically and politically there has been a strengthening of its already strong hand. Militarily and tactically it has become relatively weaker because of the growing threat and greater reality of United States IRBM and ICBM rockets, especially the submarine-launched Polaris rockets. The PVO has certainly had further re-equipment with more modern fighters and in the view of this writer further modernisation of General Klimov's PVO fighter regiments is likely to take place in the next two or three years. Already the supersonic MIG 19 is being replaced by the MIG 21 which is capable of twice the speed of sound and is being generally fitted with infra-red homer air-to-air guidance missiles. As conventional armament the MIG 21 is reported to have three 37 mm. cannon and its own airborne tracking radar for air combat in conditions of poor visibility. A supersonic successor to the heavier twin-engined YAK 25 will almost certainly enter the night-fighter all-weather regiments of the PVO command in the coming year or so, if it has not done so already. Finally Sukhoi is developing single-seater delta-wing fighters, one of which is probably already replacing the MIG 19 because of its greater speed. Certainly we must expect faster and more powerful Soviet fighter planes to appear in the early 1960s with top speeds of perhaps 2,000-3,000 m.p.h. Biryuzov's appreciation of Western air potential must surely take account of the atomic bomb carrying fighter bombers, and carrier-based aircraft as well as the heavier bombing machines of the U.S. Strategic Air Command and of the R.A.F. V-bomber force. As long as these exist in strength he is likely to retain strong forces of fighter planes.

Until a year or so ago the bulk of the PVO force consisted of over 10,000 radar-guided anti-aircraft guns and between 100 and 150 fighter air regiments with a total strength of 4,000—perhaps 5,000—aircraft. The number of conventional gun batteries is now being reduced and replaced by guided-missile ground-to-air batteries of the T 6 and T 7 pattern of which there are perhaps over 100 batteries with the concentration in the Moscow area. It is likely that the total number of purely PVO fighter regiments will soon be reduced and that there will be more dual-purpose fighters in the tactical air regiments of the "Frontal Air Armies" which will be concerned partly with the perimeter defence of the USSR in border zones.

Continual operational exercises, a high level of flying (élite so-called Class I pilots man most of Klimov's fighter air regiments) and a high standard of general morale and efficiency should make the PVO in theory one of the most effective military arms of the Soviet defence forces. In practice it is likely to be the weakest, the most vulnerable and the main reason for the Kremlin's cautious military policy of the last decade in Korea, Suez, Irak, Syria and in the support of Pekin's claim to Formosa.

In 1957 when the long-range rocket moved gradually from the realm of a potential to a reality, General Partridge, Commander-in-Chief of North American Air Defence made a statement which Marshal Biryuzov might, if he were allowed to, also make to-day and with even greater pertinence. He said: "If the aggressor's weapon is the long-range rocket, the continent stands to-day almost as naked as it did in 1946, for I have no radar to detect missiles and no defence against them." Something of the vastness of the Soviet military problem in constructing air defence against modern long-range air and rocket attack is occasionally, though naturally rather infrequently, the subject of discussion in the Soviet military press. One such article by Major-General Kornyenko referred to the need for the USSR to establish a chain of radar stations hundreds of miles outside her own territory in order to give adequate warning of an enemy air or rocket attack on Soviet territory. In this connection General Kornyenko went on to stress the importance of the signatories of the Warsaw Treaty, of whom oddly enough he singled out Hungary and Albania for special mention. But the General's article made no reference to the uncertain results which even the most modern radar equipment provides. Mr. Gromyko however did just this in a 1958 speech, when he referred to the confusion on the radar screen between meteorite blips and rocket or bomber flights. He might have added other sources of confusion such as locally fast-moving clouds, flocks of birds or the uncertain eyes of the human being needed to watch the screen and assess the flurry of dots and blips. Nor should one forget the conscious confusion created by jamming equipment carried on enemy aircraft and submarines which, unlike radio jamming, gives no aural evidence that it is in action. Soviet radar, like Western radar, is not a watertight device which guarantees adequate warning of enemy air or rocket attack. Finally airborne radar carried in bombers, now a vital part of the Soviet outer early warning system in the Arctic, Baltic and Asian regions can suffer at low temperatures, especially at heights of over 20,000 feet.

Moreover modern long-range rocket attack, e.g. by the U.S. Thor, Jupiter, Polaris or Minuteman will operate in the 10,000-20,000 m.p.h. speed range. Soviet jet fighters however fast can clearly achieve nothing against such attack and their present ground-to-air guided missiles which, incidentally, must rely initially on the Soviet radar screen, have not the range and in most cases the speed to provide adequate protection against IRBM and ICBM missiles. Even the Soviet fighters with their great supersonic speeds could be of little use against low-flying, hedge-hopping tactical bombers carrying small atomic bombs against coastal targets in, for instance, the Black Sea or Baltic Sea areas. And as for the heavier Western bombers carrying thermo-nuclear weapons, including the stand-off bomb, Soviet fighters and ground-to-air guided missiles of the T 7 and T 8 category can scarcely hope to inflict losses of over fifty per cent on the raiders which may be a minimum requirement if catastrophic damage to the USSR is to be avoided.

It remains to consider the prospects of an effective anti-missile missile and of course the important air defence potential of Soviet surprise or pre-emptive long-range attack on NATO and U.S. bases. The latter is dealt with extensively in other chapters where the conclusion is that given a moderate amount of mobility in Western strategic bombers and long-range rockets, the Kremlin can rely less and less on being able to neutralise an important proportion of Western strategic air attack by striking initially in a surprise assault with maximum resources of the Soviet DA or the other bomber and rocket commands. The problem of assessing Soviet success against U.S. aircraft carriers, which have an important atomic striking potential, is clearly beyond the Kremlin crystal-gazers and the growth of the U.S. rocket launching atomic engined submarines will make any future assessment even more problematic and hazardous.

The Soviet Union is developing an anti-missile missile with all speed. It is of interest to note recent comment on its progress given by Pokrovsky, a senior member of the growing army of top Soviet rocket specialists. He wrote: "It must be kept in mind that a missile warhead moves along a trajectory and cannot manœuvre at all. Therefore if a rocket is pinpointed in two points in succession by radar it is possible to determine the rocket's further path, the point of fall and the moment of explosion by means of electric computers in a fraction of a second. This enables anti-missile missiles to be directed automatically into the area through which the missile will pass, and to be exploded a little in advance of the path. The rocket, moving with tremendous speed and meeting these fragments in its path, will be seriously damaged. Any damage to its warhead may lead to its burning out like a meteor or to destruction through premature detonation." He went on to add cautiously: "Of course, all such measures demand a complex development of technology and tactics. However, in principle a defensive anti-missile missile is possible."

No one of course doubts the possibility and the potential. But what of the anti-missile missile as an effective or even feasible form of air defence against thermo-nuclear and atomic attack? It would appear that the advent of the long-range rocket-firing submarine, combined with the combination of ICBMs, IRBMs, and low-flying and high-flying aircraft present radar with virtually insoluble problems of detection, quite apart from its innate weaknesses already referred to. How for instance would radar pinpoint two successive stages in the path of an oncoming rocket if the frailties of the flickering screen had to deal simultaneously with several rockets fired at different ranges and therefore different trajectories from widely dispersed submarines, and at the same time land-based rockets fired from, say, Britain and the U.S.A.? The path to Leningrad and Moscow from some submarines or sites is identical; the rocket parabola is different because the range is different. The decision to fire the anti-missile missile could be made even more difficult by the use of blank rockets and all this takes no account of radio counter-measures against Soviet radar.

In short, when one looks at Soviet air defence problems, one understands the current caution in Soviet foreign policy, though no doubt the cold war and the "brinkmanship" in Berlin, Syria, Irak and other parts of the world will continue.

CHAPTER 8

LONG-RANGE AIR ATTACK

THE 1957 advent of the Soviet sputniks and the successful testing and subsequent development of Soviet intercontinental ballistic missiles especially in the past two or three years have been more than sufficient to underline the recent advances of Soviet air power in the sphere of long-range strategic air attack. In the present thermo-nuclear age of 15,000 to 20,000 m.p.h. rockets, It is easy to overlook the fact that the Soviet Air Force was the first to build and fly a substantial force of four-engined bombers. They were TB 3 machines designed by Tupolev before the end of the First Five Year Plan period (1928-32). And before Tupolev's four-engined bombers came into squadron service in the middle 1930s the Soviet Air Force was already equipped with a series of so-called heavy bomber (TB) squadrons, though they were in fact only twin-engined planes of medium range and carrying capacity called the TB 1 and TB 2.

The trends of pre-war Soviet air policy were certainly conflicting. Dr. Raymond Garthoff in his book *How Russia Makes War* makes the point that Douhet's theories of the crucial importance of long-range air attack made little impact on Kremlin military thinking. This is, of course, basically true. But nevertheless by 1936 the Soviet had formed a long-range bomber command under General Golovanov and it comprised several hundred TB 3 four-engined machines, being at the time the largest long-range heavy bomber force in the world. However, the late 1930s saw a reversal of Soviet bomber policy. It is difficult to know which of the many reasons for the reversal were the most important. Certainly the Soviet bombing in Spain and Finland of enemy communications achieved little effect largely because the attacks were not mounted in adequate strength and because standards of navigation and bomb aiming were low. But then this was also largely true of much of R.A.F. Bomber Command's attacks on Germany of the 1940-41 period; yet the relative failure did not stop the growth of the Command to a formidable force of over 1,000 four-engined bombers in the latter part of the war.

The more important factors in the decline of the Soviet long-range bomber arm in the late 1930s were these. First the Stalin purges of the 1937-38 period eliminated many of the senior military advocates, such as Marshal Tukhuchevsky and General Khripin, of a strong long-range bomber force. Then there was the failure of Tupolev to design a worthy successor to the TB 3 which had become obsolescent by the immediate pre-war period. The four-engined TB 7, though it brought Molotov to Britain in a blaze of publicity in 1942 was not a successful aircraft. In fact it was partly re-designed by Petlyakov and featured in some of the wartime Soviet long-range bomber units as the PE 8. But much more important was the pre-war need of the Soviet Air Force to expand its aircraft industry's total output and its front-line strength of fighters and fighter bombers. With the threat of war both in the East and the West the Kremlin saw that instead of a front-line strength of some 4,000-5,000 aircraft a minimum operational target figure of 8-10,000 aircraft must be aimed at. In such circumstances the emphasis was bound to be on the production of the single- and twin-engined planes such as the MIG and LAGG fighters or the Ilyushin and Petlyakov Sturmovik and medium bombers, which were more economical to manufacture in terms of man hours, raw materials and of course the number of aero-engines required. It was primarily for the same reason that the long-range transport arm of the Soviet Air Force suffered in this same period, although the expanding force of Soviet parachutist and airborne troops urgently needed a strong force of long-range transport aircraft to make it fully mobile and effective.

In any event the outbreak of the Second World War found the Soviet Air Force long-range bomber arm in a state of relative poverty. In the disastrous early months of the Germano-Soviet war, such long-range bomber aircraft as existed were thrown into the tactical air battles piecemeal, or were used for emergency transport work. Despite the tempting rail and road communications targets offered by the widely scattered Wehrmacht forces, targets which had virtually no air protection, the Soviet Air Force failed to launch any air attacks of consequence on the German Air Force or Army land or sea communications or indeed any other strategic target systems. This was in contrast to the First World War, where some of the most successful long-range bomber operations were carried out by the Imperial Russian Naval Air Service flying Russian-built four-engined Ilya Mouremetz bombers.

But the belief in the importance of strategic bombing must have been persistent in Kremlin military thinking despite many Soviet high-level pronouncements to the contrary, for in the spring of 1942 Stalin again created a long-range bomber-cum-transport force, called the ADD. These letters stand for long-range aviation (*aviatsia dalnego deistvia*), but the operations of the ADD in World War II and the nature of its wartime equipment scarcely lived up to the title and purpose of the new Air Command. It was once again commanded by General (later Marshal) Alexander Golovanov and was subordinated directly to the Soviet Defence Ministry. Throughout the war it never exceeded an operational strength of 1,000 aircraft and the bulk of these were Lend Lease North American B 25 Mitchell twin-engined medium bombers and twin-engined Soviet built Ilyushin 4 and DB 3F aircraft. The element of genuine four-engined long-range bombers—Soviet designed PE 8s—rarely exceeded ten per cent of the total operational strength. Between the spring of 1942 and the end of 1944 when the ADD operations came temporarily to an end, the great majority of ADD sorties consisted of tactical bombing of targets in or immediately behind the battlefield. The ADD also carried out thousands of air supply and troop carrying transport sorties in support of Tito's partisans, Soviet guerrilla fighters and, not least important, ferried many thousands of reinforcing crack Soviet troops to the Stalingrad front at the end of 1942, when Stalin's name town was acutely threatened by the Wehrmacht's attack.

The long-range bomber attacks of the ADD during World War II were modest both in scale and in their effect on the targets attacked. A normal bomber raid of the 1943-44 period, when the Soviet ADD was at its wartime best consisted of some 75-100 bombers of which usually only a small proportion were PE 8s; and even these four-engined bombers never carried more than five tons of bombs for a radius of action of some 600-700 miles. The navigational standards of the Soviet ADD crews were in general low. Even in some of the crack guards air regiments only half to two-thirds of the aircrews were properly trained in blind-flying wireless procedure. The absence of long-range navigational beams and radar bomber aids such as the German Knickebein and the British Oboe and H2 X, or of effective pathfinder tactics, inevitably reduced Soviet bomber efficiency at night to a low level. The absence of a long-range Soviet fighter escort prohibited the intensive use of ADD forces on long-range bombing missions in daylight. Finally, the need to use the ADD units for air tactical support and air transport missions, reduced their long-range bomber effort to a point at which it was having little or no effect on the overall military situation on the Eastern Front. It was not surprising to find that at the end of 1944 the ADD force became one of the ordinary tactical air armies—the 18th.

However remembering that from the spring of 1942 to the end of the Second World War, Anglo-American heavy bomber forces were doing exactly what a Soviet heavy bomber force might be expected to achieve in attacks on the German aircraft industry, the oil industry, the Ruhr, and other target systems, Stalin would be unlikely to duplicate this Anglo-American effort in a series of long-range air operations both costly and complicated to mount.

Even though Soviet long-range night bomber sorties never encountered the full weight of the German anti-aircraft flak guns or of General Kammhuber's 12th Air Corps of night-fighters, it is significant that from 1943 onwards the Luftwaffe began to build up a night-fighter organisation in Poland and East Prussia mainly to oppose the mounting threat of Soviet night bombers in these areas; this never amounted to more than 100 Junkers and Messerschmitt twin-engined night-fighters, i.e. some ten per cent of the total Luftwaffe night-fighter squadrons, but it was at any rate an indication that in the German Air Force view, the Soviet ADD night-bomber units were gaining in determination, experience and accuracy in their medium-scale night raids on Tilsit, Ploesti, Danzig, Konigsberg, Constanza, Budapest and Berlin. Apart from the handicap of the lack of radar bomb sights and beam navigation aids, Golovanov's long-range bomber units had no block-busting or other heavy specialised bombs. Most of the Soviet stock-pile consisted of 100, 250 and 500 kg. bombs with a small number of 1,000 kg. bombs. The ADD was indeed like a heavyweight boxer without a really accurate destructive punch throughout the Second World War. It was typically realistic on the part of the Kremlin defence authorities to use this Soviet air arm primarily as a weapon of tactical air support and air transport.

It has been stated that the Soviet authorities had a political motive as well as military reasons for not making intensive and extensive use of the ADD as a long-range bomber arm in the Germano-Soviet war. From the military point of view, as long as Roumanian oilfields, Balkan railways and targets in East Prussia, Poland, Hungary and Czechoslovakia were on the 1944-45 bomber target list of Anglo-American air forces, the ADD was better employed on tactical and air transport work from the Soviet point of view. Politically, the Moscow strategists must have been thinking intensively of post-war settlement in the last year or so of the war. They were much more conscious than the Western allies of the Clausewitz dictum that "War is nothing but a continuation of political intercourse". With this in mind they may have deliberately desisted from large-scale attacks on German cities in the final phases of the war, attacks which they could have easily carried out in 1945 with their massive short-range tactical and medium bomber air forces. This would permit the Soviet propaganda machine to accuse Britain and America of unnecessary devastation of German and other European cities and so perhaps wean the German and other nations to the Marxian Communist viewpoint. Certainly Radio Moscow output, e.g. in 1948 and 1949, included a number of broadcasts on this theme which promoted the canard.

But on the whole it was the operational deficiences, the lack of suitable long-range bombers and the deficient training rather than political cunning which limited the offensive operations of the Soviet ADD during the Second World War. Indeed when it was redesignated the 18th Air Army in December 1944 and used thereafter almost entirely in a tactical air rôle it was in effect a Soviet confession that they could not at that stage in their evolution raise and train a modern four-engined long-range bomber force.

But in the early post-war period the entire strategic air picture was changed and a Soviet re-appraisal of the importance of long-range air attack was evident. There were a number of factors in the revival and modernisation of the Soviet ADD. Of these the most important was probably the dropping of the atomic bombs on Hiroshima and Nagasaki. But, apart from this Stalin, who then tended to try to dominate Soviet military policy and strategy, was almost equally impressed by two other facts. The first was the devastating damage caused by Anglo-American bomber raids on such cities as Berlin and Leipzig which his senior air staff officers reported on in the immediate post-war period—a sort of Soviet version of the Anglo-American Bombing Survey report; the second was the revolutionary military potential of long-range rockets in the future. The relative success of the V 2 against London and Antwerp had been only moderate, but Dr. Sanger's blue-print for a strategic rocket skip bomber with intercontinental range and high supersonic speeds and the Peenemunde blue-print for a two-stage A 9/A 10 long-range rocket made Stalin conscious of the need to revise his notions on strategic air power in the post-war world as quickly as possible. As he raked over the lavish technical air inheritance he was accumulating from captured Luftwaffe material, he is reported by Soviet eyewitnesses to have asked for a rocket which would make President Truman sit up and take notice in Washington. During the Second World War both Stalin and Golovanov had tried without success to get a quantity of B 17 and B 29 four-engined bombers under the Lend Lease arrangements, which incidentally added thousands of U.S. planes to the Soviet air potential. Fortunately for the Kremlin three Boeing B 29 Superfortresses force-landed in Soviet Far Eastern Territory in 1944 in the course of long-range air operations against Japan. These were readily and forcibly annexed by the Soviet authorities and Tupolev—now their chief heavy-bomber designer, as Petlyakov was killed in an air crash about this time—was given the task of producing a Soviet version of the U.S. B 29 and was told he could have all the production priorities needed to give the USSR a powerful long-range bomber arm. It was in this period too that the Soviet acquired the Western secrets of the atomic bomb. The story of the successful Soviet espionage in this field has been told in full in the Canadian Royal Commission Report, published in June 1946, and in Igor Gouzenko's *This was My Choice* (Eyre and Spottiswoode, 1948). They throw interesting sidelights on the kind of atomic intelligence work the Soviet Union carried out against those who were then officially her allies.

In 1946 the recreated Soviet Strategic Bomber Arm was named the DA the initials standing for long-range aviation. It differed from its predecessor organisation the ADD in a number of important respects. To begin with it had a force of homogeneous Soviet-built genuine long-range four-engined bombers which were similar in performance in the 1946-48 period to some of the contemporary strategic bombers in the Royal Air Force and the United States Air Force. Tupolev's Tu 4 was externally very similar to its U.S. B 29 progenitor. The Soviet-built engines were virtual copies of the U.S.

Wright R 3350 engines, which were 2,200 h.p. radials, though the Soviet engine may have had a rather lower rating of some 2,000 h.p. However, by the time the Korean War had broken out in the summer of 1950, Tupolev's four-engined bombers were powered by Svetsov 3,000 h.p. engines. In armament Soviet 12·7 mm. guns replaced the U.S. half-inch Brownings on the B 29, and the Soviet preferred to use their own design of four-bladed airscrews.

It was certainly the 1944 legacy of B 29s which helped the Soviet DA to have several hundred Tu 4s in their long-range bomber arm by 1950. The Soviet Tu 4 bomber was in series production at factories in or near Moscow, Kazan, Tashkent and Novosibirsk by the end of 1946. But the Tu 4 had a practical radius of action of only 1,500-2,000 miles. In terms of global strategy in the atomic age it was unworthy of the stock-pile of Soviet atomic bombs which by 1950 was being steadily amassed. The vulnerability of the B 29 to Chinese-flown MIG 15 fighters in the Korean War was an indication of the kind of difficult operating conditions which the Tu 4 might encounter in daylight operations against better trained U.S. or British jet fighter pilots, backed by an efficient early warning system. And at night or in bad weather the Tu 4 crews still lacked the friendly help of radar bomb sights and long-range navigation aids.

But the creation and expansion of the long-range bomber DA force was not the only aspect of Soviet progress in strategic air attack weapons during the 1945-50 period. The Soviet Army and Navy now came into the field and a Russian improved version of the German V 2 rocket went into a series of test-firing trials, including submarine-launched trials. The latter were witnessed by a number of Scandinavian journalists from merchant ships operating in the Baltic in 1949-50. However, what was headline news in Scandinavia for a day or so scarcely rated a mention in the Western press. This was the period before the Korean War when any recognition of Soviet technical air progress might earn for a writer the label of left-wing intellectualism.

Throughout the 1950s there has been continual evidence of Soviet progress in all the important aspects of strategic air attack. By 1953, the DA force of Tu 4 bombers had expanded to three full air armies, with an operational strength of about 1,000 planes according to General Twining, then Chief of Staff of the U.S. Air Force. It was in August of that year that Malenkov told the supreme Soviet that "the United States no longer has the monopoly in the production of the hydrogen bomb" and four days later his words were given substance when a thermo-nuclear explosion was detected on Soviet territory. And it was also in 1953, that Gordon Dean, Chairman of the U.S. Atomic Energy Commission since 1950, propounded the view in his *Report on the Atom* that the USSR had all the main ingredients of a powerful atomic energy programme, viz., adequate material resources which included enough uranium, the scientific competence, the technological and production capacity and finally the will, determination and ability to use her atomic energy resources. Mr. Dean was one of the first to draw public attention to the great post-war expansion in Soviet technological training, which together with the increased production of such things as steel, aluminium, electric power and petroleum, gave the USSR the means of "the making of an ominous war machine geared to the manufacture and use of either conventional or atomic weapons". Mr. Dean estimated that about 30,000 Soviet engineers had graduated in 1952 and added, "the figure may increase to about 40,000 by 1955". This prophecy was apparently fulfilled.

And the West did not have to wait long for further concrete evidence of Soviet progress in the field of strategic air attack. By 1953 Golovanov had faded from the field of Soviet strategic air power and had been replaced by Marshal of Aviation Sudetz who had one of the best records as an operational commander of tactical air forces in the Second World War. He had previously been Chief of Staff of the Tactical Air Forces under Marshal of Aviation Konstantin A. Vershinin. He was to press the Soviet jet bomber designers such as Myasischev, Ilyushin and Tupolev and jet-engine designers Kutsnetsov and Mikulin to close the gap which still existed between the long-range jet forces of the USAF strategic bomber arm and the conventional piston-engined Tupolev aircraft of the Soviet DA.

Sudetz's long-range air command was directly subordinate to the Soviet Defence Ministry. His energy, influence and organising ability, accompanied by an unusual spurt in the design of powerful jet engines by the Mikulin and Kutsnetsov design bureaux, were soon to end the situation of 1950-55 during which Soviet strategic jet bomber aircraft lagged badly behind those of the United States. The development of Mikulin's large AM 3 turbo-jet which was raised in power to produce a static thrust of some 15,000 lb. helped to close the technical gap. By the end of 1954, the Tu 16 twin-jet medium bomber powered with two of these large jet engines began to enter squadron service with the Soviet long-range bomber arm. The new Tupolev jet bomber, despite having only two engines compared fairly well in performance with the Boeing B 47 six-engined Stratojet which, however, had entered U.S.A.F. service two or three years previously, was available in much larger numbers and had the advantage of advanced bases in Europe, Asia and North Africa which permitted the B 47 squadron to effect a much deeper penetration of Soviet metropolitan territory than the Tu 16 could hope to achieve against targets in the United States. The Tupolev 16 had a maximum range of some 4,000-4,250 miles, that is a radius of action of only 2,000 miles. It had a good top speed of approximately 550 m.p.h. and, with the help of flight refuelling, clearly constituted a new Soviet air threat to the North American continent. But a much more important 1954 jet bomber product of the Soviet aircraft industry was the four-jet Myasischev "*Bison*" 550-m.p.h. bomber which could carry atomic or thermo-nuclear bombs to targets up to 3,000 miles from its base without flight refuelling. And in 1955 came the news of a large longer-ranged but slower bomber, the Tupolev 20, powered by four 12,000 h.p. Kutsnetsov turbo-prop. engines; this new bomber had a radius of action of some 3,500 miles, which with the help of flight refuelling could reach virtually any target in the North American continent. The two Soviet four-engined jet bombers began to replace the obsolescent Tu 4 piston-engined aircraft of Sudetz's DA Strategic Air Command during the 1955-56 period, and re-equipment with these jet bombers continued throughout the next four or five years. To-day (1960) the air regiments of the Soviet long-range bomber force are nearly all equipped with twin- or four-engined jet bombers.

But much more important in the long run was the Soviet post-war development of Soviet long-range rockets with atomic warheads in a wide range of categories from the single-stage tactical T1 with an initial range of some 300-500 miles to the intercontinental ballistic missile such as the T 3 of over 5,000 miles range. It is often assumed that Soviet progress in the production of long-range military rockets in the 1950s was due almost entirely to the German rocket legacy. Certainly the inheritance from Peenemunde and other German long-range rocket centres in Poland was in large measure responsible for the accelerated rate of Soviet progress in these weapons in the 1945-55 period. But there had been a long tradition of Russian practical study and encouragement of ideas in this military field, going back long before the October Revolution of 1917 and continuing under the Soviet régime. Konstantin Tsiolkowsky, the pioneer of Russian long-range rockets, produced the blue-print of a military rocket using the combination of hydrogen and oxygen as liquid-fuel propellants as far back as 1903. In 1915 he conceived a larger rocket which used a combination of hydrocarbon and oxygen fuels. On the eve of the Bolshevik Revolution he was designing a long-range three-stage rocket. Tsiolkowsky began to assemble a team of Soviet rocket scientists and engineers including Rynin, Perelman, Zander, Tichhof, Seylinger, Scherchevski and Wetchinkin. In the 1920s and early 1930s they carried out valuable design and study work at TSAGI rocket research centres near Leningrad, Moscow and Kazan. When Tsiolkowsky died in 1936, he had established a rocket tradition of research and development in both large liquid-fuel rockets and small solid propellant rockets. He had brought the mathematical genius of Hermann Oberth, the great pre-war expert on rockets and space travel, into the Soviet rocket literature, tradition and orbit. He had done much to lay the foundation which enabled the Soviet to establish a useful corps of rocket scientists and engineers who could exploit fully the advanced German rocket weapons and blue-prints which fell into the Soviet hands in 1945-46.

However, during World War II, the Kremlin had to concentrate on tactical weapons for the ground battles, and so rocket engines and missiles were developed to that end and with at least as much success as in most other contemporary armed forces. Thus by the winter of 1941-42, Soviet factories were making large numbers of powder rockets called Katyushkas which made considerable impression on the Wehrmacht, especially in the battle for Stalingrad during the winter that followed. The Soviet Air Force developed air-to-ground rockets fired from fighter and Sturmovik aircraft before any other air force in the world. In may be recalled that in 1941 Lord Beaverbrook was anxious to acquire some of them from the USSR as a form of Lend Lease in reverse. However, to judge from German military appreciations, these air launched tactical short-range rockets were not particularly effective when used against German armoured or other forces on the Eastern Front during the Germano-Soviet War.

Certainly in the decade that followed World War II, the Kremlin paid much greater attention and gave higher priority than did the Western powers to the development of the German long-range rocket legacy. In the first half of this period, from 1945-50, reports of Soviet developments of German V 2 rockets and of the German A 9 and 10 blue-print for a two-stage rocket were generally discounted or underestimated in Western military circles. When, for instance, Dr. Tellman an ex-Wehrmacht rocket specialist left the USSR in 1949 and went to the Argentine to join the post-war colony of German aircraft and rocket specialists who had gathered there, his accounts of Soviet developments in both single- and two-stange long-range rockets were generally ignored. And yet his views were supported by other German sources who pointed to such things as the work being carried out at Khimki near Moscow and at Ostakhosh further east to improve the range and guidance of the German V 2 rockets.

And by the 1952-53 period there was repeated evidence of an improved Soviet version of the German V 2 going into large-scale production, with liquid fuel rockets, weighing some 12½ tons and with a range of up to 500 miles. Dr. Walter Dornburgher, who was the general in command of the Peenemunde long-range missile development centre during the Second World War and who is now better known for his work in the United States with Dr. Wernher von Braun on U.S. Army long-range rockets, has perhaps the greatest knowledge of the Soviet long-range rocket developments during the period when German influence was paramount, i.e. from 1946 to 1955. He has certainly been able to have contact with many of the German rocket scientists and engineers who worked in the USSR during this period. These ex-Wehrmacht rocket specialists reported that by 1950 the Soviet had developed the V 2 to the point where it would fire to a range of 300 miles and with greatly improved accuracy—some German sources said to within 500-1,000 yards of the target aimed at.

Nevertheless these rocket developments did nothing to solve the problem of Soviet strategic long-range air attack. And so up to 1955 the absence of an efficient two-stage Soviet rocket and of a force of long-range modern jet bombers, kept the Soviet Union well behind the United States in the struggle for supremacy in the crucial field of strategic air power.

It was the next five years (1955-60) which were to see most of the developments which gave the Soviet Union broad parity with the United States in the crucial field of strategic air attack, though clearly the technological progress of the early 1950s had laid the foundations of this parity. It was for instance then that the Soviet Union began to stock-pile both atomic and thermo-nuclear bombs, to build its native designed Mikulin and Kutsnetsov engines powerful enough for the new long-range jet bombers which were coming into service and to test-fire the T 2 intermediate-range two-stage rockets. It was also between 1950 and 1955 that the techniques of long-range flight refuelling were worked out and Arctic air and rocket bases extended. Both these latter developments added potential range to the Soviet long-range air attack against targets in North America. Indeed the majority of the air regiments of the Soviet DA jet bomber force today have advanced bases in the Arctic area, and an advanced headquarters near Magadan commanded by General Kapitsa while General Sudetz commands the whole of the DA long-range bomber force from his Moscow headquarters. Between 30 and 40 new jet bomber bases are reported to have been constructed in the Arctic regions of the USSR during the last five years. These incidentally have included some built after the successful test-firing of the Soviet intercontinental ballistic missile in August 1957. New long-range jet bomber bases have been built or expanded in the 1955-60 period on the islands of Severenaya Zemlya, Novaya Zemlya and Franz Josef Land and along the Arctic mainland from Murmansk in the West to Anadyr in eastern Siberia.

But the bulk of the Soviet DA air regiments have their permanent air bases near Moscow, Kiev, in the Carpathians and the Far Eastern Maritime Provinces, where the problems of servicing, operational training and the testing of new equipment are simpler to handle than in the Arctic north. During the past two or three years, despite the Soviet advances in long-range rockets, the modernisation and re-equipment of the long-range bomber arm have continued. There has been refitting with the new turbine-powered jet bombers the Bison and the Badger, so that to-day (1960) the Soviet DA is virtually an all-jet air force, with the piston-engined Tu 4 almost eliminated from operational bomber duties and diverted to tanker refuelling sorties, advanced bomber training and maritime reconnaissance work with the Soviet naval air forces (the A-VMF). The slower Tu 20

Bear four-jet turbo-propeller bomber is likely to be phased out of the DA bomber force to do the same work as that allotted to the Tu 4, especially when the new Soviet supersonic delta-wing jet bomber comes into general service with the DA as it is expected to in fairly large numbers in the early 1960s. The new bomber has been dubbed the Bounder by the NATO authorities and is credited by some sources with an intercontinental radius of action of 5,000 miles and a stop speed of about Mach 2. It is interesting that it was reported to be going into service in DA jet bomber units early in 1959 some eighteen months after the public utterances of Khrushchev and Marshal of Aviation Vershinin, C.-in-C. of the Soviet Air Force, to the effect that Soviet ballistic missiles were making bombers obsolete. This is a sharp reminder of the hiatus between Soviet declarations and Soviet military policy. "Fighter and bomber aircraft can now be put into museums" declared Khrushchev in the autumn of 1957. But since then re-equipment of the Soviet Air Force with faster supersonic fighters, fighter and tactical bombers and strategic intercontinental bombers has continued. The Soviet DA force continues to flourish and to constitute currently one of the main weapons of Soviet long-range air attack on North America. If the Soviet defence authorities really believed what they say about the lack of future for manned strategic bombers it is hard to explain the continuously intensive operational training of the long-range bomber force, including the intensive flight refuelling operations and the day and night blind-flying training. Then there is the recent improvement of the long-range beam and other navigational aids and of the DA aircraft radar bomb sights. Both Badger and Bison jet-bomber units have been refitted with longer fuselage noses housing new radar bomb sights within the last two or three years. Nor should one ignore reports that the Soviet Air Force are developing a new long-range atomic engined transport bomber as well as an intercontinental manned rocket bomber, quite apart from the T 4a unmanned rocket bomber.

It is therefore hard to see any likelihood of a drastic reduction in the operational strength of the Soviet long-range bomber force in the next two or three years. It will probably consist of some 600 twin-jet Tu 16 bombers, perhaps 500-600 force jet Bisons Il 38s, between 100 and 200 turbo-propeller Tu 20 Bears and perhaps 50-100 jet delta-wing supersonic Bounders. It is likely to be the most important element in the Soviet ability to attack all areas of the United States, though in the middle and late 1960s this position will no doubt be changed as Soviet rockets with intercontinental range improve their accuracy and reliability and increase their numbers, and as the Soviet Navy develops a useful force of ocean-going submarines capable of firing long-range Golem and Komet rockets.

It is very difficult to calculate the moment at which the Soviet long-range rocket force will replace the DA as the main element in the strategic air attack potential of the USSR. One thing is clear from all the rocket intelligence which has emerged in the last five years. Soviet rocket plans and achievements for the period 1955-60 have surpassed those of the United States. This is partly because the Soviet High Command made better use of the German long-range rocket legacies and gave immediate post-war rocket plans a higher priority than did the NATO countries. In addition the Soviet Union has concentrated on fewer types of long-range offensive missiles; they have not been plagued by the inter-service rivalries which have slowed up the United States long-range missile programme. The present (1960) gap between the United States and the USSR is tentatively estimated at a year to eighteen months in favour of the Soviet. In any event a breakthrough in guidance, in greater stability of the liquid-fuel propellants, in developing atomic rocket engines or in large solid-fuel rocket motors could change the whole picture in a year. At the moment both the USSR and the U.S.A. rely on liquid-fuel long-range rockets. Since 1956 the USSR has been manufacturing the T 2 intermediate-range rocket capable of carrying an atomic warhead a distance of 1,750-2,000 miles, and since 1958 the T 3 and T 3A intercontinental ballistic missile each capable of carrying a thermo-nuclear hydrogen warhead. The T 3 has a reported range of 5,000 miles and the T 3A an even greater range of over 6,000 miles, enabling it to reach any target in North America from its present Soviet bases. But one needs with each year that passes to assess and reassess both the reliability and accuracy of long-range missiles. No one can doubt that both sides have built rocket motors powerful enough to carry the atomic and thermo-nuclear warheads the distances required. There is every reason to doubt the current reliability of the Thors, Jupiters Atlases and Titans on the U.S. side and the T 2, T 3 and T 3A on the Soviet side. We know of course more about U.S. rocket and aircraft failures than of Soviet mishaps. It is only when Soviet aircraft carry foreign passengers that the news of their air crashes have been publicised, but the news of them has leaked out to Western intelligence sources. As for U.S. long-range rockets, the engines or guidance malfunctioned in roughly fifty per cent of reported firings in the 1957-59 period and there is no reason to believe that Soviet rocket performance is either better or worse, or that the USSR will eliminate unreliability and instability, in liquid-fuel rockets before the United States or perfect the elements of guidance and re-entrance from stratosphere heights to the earth's surface.

At the moment the Soviet Union's main numerical long-range rocket advantage lies in the T 2 intermediate-range rocket capable of firing distances up to 1,500-2,000 miles or so. At the time when the first Soviet test firings of the T 2 were being carried out (1954-55) the U.S. Army Redstone rocket was being limited in its potential range to only 200 miles, as an act of deliberate policy. Incidentally this U.S. Army rocket, designed under the team leadership of Werner von Braun, the famous German V 2 rocket designer, shared one of the basic weaknesses of many if not all Soviet and U.S. rockets. Prior to firing, the liquid oxygen tanks have to be continuously topped up to make good evaporation losses. This imposes a firing delay. In August 1959 the British Ministry of Defence announced that the British based Thors could be fired in fifteen to twenty minutes. During this period enemy rockets could travel several thousands of miles and so deter the deterrent. The interval for firing published to impress was in fact a confession of a dangerously slow firing time.

But despite the disadvantages of liquid-fuel rockets and their uncertainties, the lowest estimates of current Soviet T 2 operational strength would put the figure at 1,000 of these rockets. Indeed some German sources think they are being produced at the rate of thousands per annum. Clearly a mass atomic attack by them on Britain or Turkey could seriously reduce the deterrent effect of NATO rocket and bomber forces based in these countries. The first versions of the T 2 had a speed of some 5,000 m.p.h. and later versions, including a much improved experimental version called the T 4, had speeds of up to 8-10,000 m.p.h. Defence by fighter aircraft or even defensive guided missiles such as the British Thunderbird and Bloodhound or the U.S. Nike or Bomarc would be quite ineffective against such Soviet rocket attacks.

What about the Soviet intercontinental ballistic missiles, the T 3 and T 3A, which probably went into production in 1958-59? Most U.S. intelligence and other sources would put their current strength at between 100 and 200 and at a total of 500-600 in the next two or three years. Do these figures help us to assess the strategic rocket potential of the USSR? Most if not all of the Soviet ICBM rockets will be of the liquid-fuel variety and subject to the same instability and unreliability to which that weapon is currently heir. The T 3 is said to have a range of about 5,000 miles and the T 3A of 6-7,000 miles. The speeds are of the order of 15-16,000 m.p.h. The rocket motors are similar to those which powered the first Soviet sputniks. The guidance system is inertial and reasonably reliable. There are however two different current estimates of the degree of accuracy of the Soviet ICBM rockets. The Soviet version published in the autumn of 1957 claimed an error of plus or minus about five miles, and the German estimate published in 1958 thought the three-stage rockets might have a tolerance of plus or minus ten miles. Both estimates of course constitute a high degree of accuracy. But the nub of the problem was given in a statement by General Pokrovsky in the autumn of 1956 when he said: "Long-range rockets can only destroy targets the co-ordinates of which are known in advance. They cannot destroy mobile targets. For this self-seeking guided missiles or piloted weapons are needed." In 1957, Pokrovsky, who is a leading Soviet rocket expert, added that the future employment of ballistic missiles would be against "targets previously known and precisely located on the map".

While large cities such as New York, Moscow and London are clearly vulnerable to long-range rocket attack, the USSR must give inevitable priority to attacking the bases of the U.S. Thor, Jupiter, Titan and Atlas rockets and of the SAC and V bomber bases. Thus the importance of making these Western forces as mobile and as widely dispersed as possible could be a matter of victory or defeat for the West. It is hard to see why the wide-open spaces of Canada and Australia are not more widely used for these purposes, despite the political difficulties and natural susceptibilities involved. The location of the bases which could be newly constituted could be hidden from the USSR. With the use of large transport aircraft and helicopters the mobility of Western rockets and bomber aircraft could present a virtually insoluble problem to Soviet planners responsible for long-range bomber and rocket attack; the preemptive Soviet blow about which so much has been written could become an empty blow, hesitant and based on out-of-date intelligence and so virtually useless.

But in the next few years, the emergence of solid-fuel rockets of intercontinental range and of a growing fleet of Soviet ocean going submarines capable of firing long-range rockets could change the entire situation. These changes are considered in other chapters of the book. But in 1960 the position is that the Soviet DA bomber force remains the chief weapon immediately threatening the North American continent and therefore the defence of the free world. In an article in a Soviet Air Force journal, General Tatarchenko wrote in 1946, "Rockets will not eliminate the necessity for any class of aircraft, least of all high speed giant aircraft, both bomber and transport." The evidence of the next fourteen years of Soviet aircraft and rocket development has borne out this prediction.

CHAPTER 9

NAVAL AIR POWER

AT the Twentieth Communist Party Congress, held in Moscow in February 1956, Marshal Zhukov, then Soviet Minister of Defence, said: "In a future war the struggle at sea will be of greater importance than it was in the last war." In the three or four years that have elapsed since Zhukov's statement, while he has fallen from Soviet grace, the point of his remarks has been sharpened. The ocean-going atomic-engined long-range rocket-firing submarine has become the latest and perhaps the most important weapon of strategic warfare. Indeed it may well be the ultimate weapon of long-range warfare on this planet.

The naval air forces of the USSR have not attained the numbers, power or prestige of U.S. naval air forces. This is partly because the Navy was for so long traditionally the Cinderella of the Soviet armed forces. And so, not surprisingly, Soviet naval air forces played only minor regional tactical rôles in the Soviet operations of the Second World War. The main reasons for this is that until 1945 land power was the dominating influence in Soviet military thinking. And who can blame the Kremlin? For surely the great heartland of Eurasia, which contributed so much to Napoleon's defeat, did much the same to Hitler. Until the end of the Second World War the Soviet Navy was probably considered in the Kremlin to be primarily an extension of the Army's land flank, a sort of coast-guard or perhaps a sentinel for Russia's seas which are largely backed by land.

This tactical conception has not always been the tradition of Russian naval air forces. In the First World War, while the bulk of naval air operations consisted of tactical reconnaissance flights, mainly along the Baltic coast, for which, incidentally, a fairly large number of U.S.-built Curtiss seaplanes were used, the Russian naval air arm was also one of the first in the world to use squadrons of native-designed four-engined bombers. These were Ilya Mouremetz machines, at first powered by French-designed and built 600 h.p. engines.

It is interesting to recall that in the final year of operations, prior to the outbreak of the October 1917 Revolution, the Russian naval air squadrons flying on the Baltic front operated as much against German army as against German navy targets. This was also to be the pattern of operations of the new Soviet naval air arm in the Second World War. It was to be a coastal air command as much as a naval air command. In other words it was an integral part of the tactical air forces of the USSR which happened to be based on airfields near coasts and seaports. Its aircraft were, in general, not of specially naval air design, though it has always had a complement of floatplanes or flying-boats from 1917 to the present day, and has carried out such special naval air operations as torpedo bomber and mine-laying attacks. But there was no intention of making it a naval air force, capable of concentrating entirely on air support for the Soviet Navy at the expense of coastal tactical air support for Soviet Army operations. From its earliest years right up to the end of the Second World War, the Soviet naval air units were in some ways similar to the naval air forces of the Wehrmacht. The Soviet Union like Germany employed no aircraft carriers, though Germany did build one, the *Graf Zeppelin*, which was launched in 1938 and underwent operational trials using a combination of dive-bombers and single-seater fighters. But no operational use of it was made in World War II and the temporary naval air squadrons of Ju 87 and Me 109 aircraft soon reverted to their old Luftwaffe tactical army support air corps in 1940. The Kremlin like the Wehrmacht saw its naval air forces as a mixture of coastal float-plane and flying-boat reconnaissance and patrol units, with the land-based bomber elements intended for both naval co-operation work such as mine-laying and torpedo-carrying but also as a tactical bomber force.

The main difference between the German and Soviet conceptions of naval air power was that the Soviet squadrons were more closely associated with the Soviet Navy and for longer periods than were the German naval air units. Finally, unlike Luftwaffe squadrons, a high proportion of the Soviet naval air regiments included land-based fighters of the same design and construction as those of the Soviet army tactical air forces. The only difference in practice was that the naval air fighter squadrons were engaged in the air defence of ports and coastal areas. The U.S. or Japanese conception of naval air power with strong aircraft carrier forces which could decisively influence the balance of sea-power over the oceans of the world has never been a Soviet conception. In view of the inevitable trend of Soviet strategy up to 1945, in which a largely continental land power was bound to use largely continental tactical air forces, the Soviet conception of naval air forces was until then appropriate for the defence tasks in hand.

The Revolution had left the naval air fortunes of the USSR at a nadir of inadequacy. The limited air power that was then available to the new Soviet leaders was bound to concentrate on the tactical support of the Red Army. Nevertheless a force of floatplane reconnaissance aircraft was gradually formed. These were designed mainly by Grigorovich but included some of both U.S. and Italian origin. By 1925 Soviet naval air forces had reached an operational strength of some 300-400 floatplanes and flying-boats. By the early 1930s it had become clear to the Soviet authorities that re-equipment with land-based fighters and bombers would make the naval air squadrons more flexible and more efficient. The USSR must be given credit for becoming aware before the Japanese and Germans of the inferior performance of floatplanes and flying-boats in modern air operations. While Germany hung on to her Heinkel, Arado and Dornier machines of this type throughout the 1930s, the Soviet Union had by 1935 decided that only a very small proportion of her naval air arm should be equipped with float-planes or flying-boats. No doubt the prevalence of ice-bound seas around the USSR was a major factor in accelerating the Soviet decision.

At all events in the four or five years prior to the outbreak of the Germano-Soviet War in June 1941, the strength of the Soviet naval air arm ranged from 600 to 800 aircraft of which some three-quarters were land-based bombers and fighters. They were in no way different from the I15, I153 and I16 fighters and the SB 1, SB 2 and PE 2 bombers in the Soviet Army Air Force. The operational training was of course different and so were the uniforms which the aircrews wore. But it may be significant that the ranks in the naval air squadrons were similar to those in the Soviet Army, ranging from "General-maior", Polkovnik (Colonel), Podpolkovnik (Lieut-Colonel), Maior (Major), Kapitan, down to Leutnant amongst the officer ranks. This practice has remained to the present day as it has in other Soviet Air Force squadrons; and it may be a psychological factor in the continuing domination of the Soviet Army in defence matters in an era in which the growing importance of rockets and guided missiles has caused some inter-service rivalry in the USSR where such rivalries are rare.

Throughout the Second World War, the Soviet naval air force underwent both modernisation and expansion but remained essentially a land-based air force with local tactical objectives. One of the most intimate but rather localised pictures of Soviet naval air operations in the Second World War derives from the co-operation of the Royal Air Force 151 Fighter Wing with them. This R.A.F. wing of Hurricanes spent a few months from September to December 1941 on Soviet naval air bases in the Arctic Circle near Murmansk, teaching ground and air staff to assemble, service and fly the British fighter planes. At the Keg-Ostrov air base they were able to observe the day-to-day standards, methods and routine of Soviet naval air units commanded by Major Kuznetsov. (The Commander-in-Chief of all the Soviet Naval Air Forces at the time was a Major-General Kuznetsov and the Commander-in-Chief of the Soviet Navy was an Admiral Kuznetsov.) The Soviet naval air squadrons on this Arctic air front flew PE twin-engined bombers and had the I 16 low-wing monoplane single-seater fighter to escort them. It is interesting to note that during the three months of their co-operation with the R.A.F. during which British pilots flew free-lance patrols and fighter escort to Soviet bombers, virtually no anti-shipping operations were carried out or indeed planned by these Soviet naval air units, despite the obvious German targets plying in Norwegian waters. The diary of day-to-day operations (see *R.A.F. in Russia* by Hubert Griffith, Hammond and Hammond, 1943) reads almost like any non-naval tactical air force command which sends its planes out on bomber interception, air patrols and attacks on enemy-held airfields. There was no sign of mine-laying or torpedo bomber operations or of regular Soviet naval air reconnaissance of enemy shipping in Arctic or Norwegian waters.

While it would be absurd to assess the general standards of Soviet naval air operations and personnel by the limited local experience of a group of R.A.F. officers and men on one or two air bases, their professional on-the-spot impressions have at least the ring of authenticity. On the whole they found Soviet naval air pilots good, experienced, interested in new tactics and able to convert to Hurricanes quickly and fly them in combat skilfully. The Soviet pilots had courage and determination despite the handicaps of flying obsolete biplane 250 m.p.h. I 153 fighters against the 350 m.p.h. German Me 109 fighter squadrons. Their technical knowledge and servicing standards were thought to be good, their anti-aircraft fire intense, speedy and accurate, their ground camouflage very good, their observer corps early-warning work accurate and rapid and their general standards of discipline and cleanliness high. One thing which particularly struck the British air contingent and other British air observers in the USSR in this period was the speed of Soviet airfield construction. Marshes were drained, roads and runways built in a time which would be regarded as impossibly short in other countries. This view of Soviet naval air standards is not necessarily typical, for no doubt the Soviet naval air units on the Murmansk front were told to put on their best bib and tucker during the R.A.F. tenancy of their air base. But high flying and technical standards cannot be attained and maintained during a three months' visit unless the basic skills and training are already there and on a large scale.

Nevertheless, as an offensive or fleet co-operation force, the Soviet naval air units on these Northern Norway-Southern Arctic fronts did not have a very substantial operational record during the Second World War. This was particularly noticeable during the passage of the Anglo-American P.Q. convoys in 1942 and 1943 from Scottish waters to Murmansk, convoys which carried tanks and other war equipment which Admiral Kuznetsov's naval air units on this front should surely have considered a major operational responsibility. These Anglo-American convoys might well have had the benefit of Soviet anti-submarine patrols and Soviet air attacks on the German torpedo bomber bases at Banak and Tromsö in Northern Norway. Although some desultory Soviet naval air attacks were made on these key Luftwaffe air bases, which were incidentally weakly defended by a handful of inexperienced Me 109 squadrons based in the area, no really sustained determined series of Soviet bomber attacks was carried out, even during the crucial period of the passage of the above-mentioned convoys. Nor did the Soviet naval authorities institute regular anti-submarine naval patrols in the area. There were no doubt at least two basic reasons for this. The Soviet authorities did not think of naval air power as a factor in protecting ships coming from long distances overseas in time of war, for overseas trade had been such an insignificant part of their activities in time of peace. But their aircrews on the Murmansk Northern Norway naval air front had neither the aircraft nor the blind-flying training and equipment to enable them to do long-range anti-submarine patrols. However, Soviet naval air fighters did provide much needed fighter escort, flying Hurricane and LAGG fighters over the Anglo-American convoys in 1942 and 1943 as they approached Murmansk and Archangel carrying the Lend Lease supplies for the Soviet armed forces.

On other areas of the Eastern Front in the Second World War the operations of the expanding Soviet naval air arm had the inevitable varying success. In the first rush of the German advance to the Eastern Baltic and towards the Black Sea, the Soviet naval air forces were used almost entirely as part of the extreme northern and southern wing in the tactical air battles of the main land fronts. On the northern wing their operations over the Baltic were ineffective, and inevitably so, during the summer months of 1941. But in the subsequent sturdy air defence of Leningrad in the 1942-43 period, naval air fighter regiments played a more than useful rôle in the protection of the city and its port facilities. As the Soviet Air forces went over to the offensive in the next two years, the naval air forces on this front were expanded and equipped with more modern Lavochkin and MIG fighters and Ilyushin and Petlyakov twin-engined bombers. In a series of mine-laying and bomber operations they harried German shipping movements in the Eastern Baltic especially during the last year of the war. Here again they failed to pull their full naval air weight in the allied cause. The central and Eastern Baltic was for example a major centre of operational training for the land-based expanding Ju 88 torpedo-bomber arm of the Luftwaffe as well as for a few bomber squadrons carrying the new German guided anti-shipping bombs, the HS 293 and the FX 1400. Had the Soviet naval air forces had the striking power and skills in the Baltic naval air command they could have carried out a series of attacks on the Luftwaffe training bases between the island of Rügen and Danzig which could have harassed and might well have reduced the expansion of the Luftwaffe's torpedo-bomber and guided-bomb units. But no such attacks were carried out, partly because the Soviet naval air units clung in the main to their tactical local rôle of defending ports and fighting in support of the Red Army in coastal operations and partly because they were not in general willing to co-ordinate their air policies with their British and American allies to exercise the maximum joint pressure on Germany's waning air resources.

It was in the Black Sea area that the Soviet naval air arm had its most striking successes in the Second World War. Soviet naval torpedo bombers, chiefly twin-engined IL 4 machines, carrying an 18-inch torpedo, or alternatively a mine weighing about one ton, made successful attacks on German shipping plying from Roumania to the Crimea or laid mines in German-held Black Sea ports; and the naval fighter regiments now re-equipped with LA 5 and YAK 3 single-seater fighters, with a top speed respectively of about 375 m.p.h. and 355 m.p.h., successfully defended Soviet-held ports such as Tuapse, Kerch, Sevastopol, Poti and Batum from the admittedly low-scale air attacks of the German 4th Air Fleet. And in the summer of 1944, flying with the Soviet 8th Air Army under Generals Ermachenkov and Tupikov, a force of some 500 Soviet naval aircraft played an important air support rôle which helped General Tolbukhin's Army Group to throw the German and Roumanian forces out of the Crimea. In the same period forces of between 300 and 400 naval aircraft on the Baltic front were thrown in to support the Soviet offensive against Vitebsk and Orscha.

The most disappointing performances of the Soviet naval air forces on the Second World War were probably on the Southern Arctic front. The intermittent raids on the German torpedo-bomber bases in Northern Norway had little apparent result. The attempt to bomb the German battleship Tirpitz when it was anchored in Alten Fjord was a failure. Only three or four aircraft located the target and no hits were registered. In the absence of beam navigational aids, radar bomb sights and blind-flying skills, the failure was not surprising.

The greatest hiatus in the experience of Soviet naval air forces in World War II was in anti-submarine warfare. No doubt the Defence Ministry felt that the protection of Soviet shipping from German U-boat attack did not constitute a key factor in the overall strategy of the USSR. Moreover to have achieved any major successes in this field would have involved them in long-range patrols, sea-mining and bomber attack and the fact is that they had little suitable modern naval aircraft and equipment with which to undertake such rôles. They ended the war with obsolescent or obsolete flying-boats mainly of American origin or design, such as the PBY 6A twin-engined patrol bomber or the earlier PBY 5 machine which was built under licence in the Soviet Union. The flying-boats and floatplanes of the Soviet Naval Air Force did not make anything like the design progress of the land-based aircraft in the Second World War. In fact virtually no new types of aircraft or equipment were produced. The aircraft industry concentrated on the more important job of expanding and improving the production of fighter planes and tactical bombers to support the Red Army.

As far as available information can guide us, the Soviet naval air forces had an operational strength of some 2,500-3,000 aircraft at the end of the Second World War of which over four-fifths consisted of land-based fighters, single- and twin-engined bombers. They played a very secondary rôle in that war and produced little or nothing that was original or noteworthy. The naval air component of the Far Eastern Pacific Command had a little operational experience in the 1945 mopping-up operations against Japanese forces in Manchuria but was otherwise mainly a source of reinforcement for the Soviet naval air forces on the Black Sea and Baltic air fronts in the Second World War.

In the immediate post-war period the Soviet naval air arm continued its rather unimportant existence alongside the main Soviet Air Force and was directly subordinate to the Defence Ministry. The testing of Soviet Naval Air Force planes was largely in the hands of the Soviet Air Force authorities. It was for this reason that its fighter and bomber squadrons were re-equipped with jet planes much later than the main body of the Soviet Air Force. While these were rapidly re-equipped with Yak 15 and Mig 9 jet-fighter machines and later with Yak 23 and Mig 15 jets, the Soviet naval air regiments tended to keep their conventional piston-engined aircraft for much longer periods in the late 1940s.

In the early 1950s there was a radical change. The Soviet naval air forces were put under the direct control of the Soviet Navy and were to be given much greater status in Soviet defence planning in the next five or ten years. There are different views as to why this happened when it did. Those who tend to see a political leitmotif in every major change in the Soviet military organisation might say that this was due to the personal esteem which Admiral Kuznetsov enjoyed in Stalin's entourage. But probably much more important was the general expansion and increased importance of the Soviet Navy as a whole. With an increase in submarine strength, especially ocean-going submarines and an increase in cruisers, there was clearly a need to bring the Soviet naval air arm up to date, to equip it with jet planes, to expand both its land-based and shipborne strength and to enable it to meet the new strategic global commitments which the Soviet armed forces now had to assume. When in 1955 Admiral Vladimirsky of the Soviet Navy said, "The emergence of new weapons and technology requires us to examine carefully anew the rôle of the navy," he was both forecasting the shaping of things to come and resuming what had already happened in the early 1950s. On Soviet Air Force Day in the summer of 1951, naval air force jet fighters and bombers (MiG 15s and IL 28s) took part in the annual Air Force flying display for the first time for some years. But perhaps of greater significance was the evidence of the 1950-51 period that long-range rockets were being test-fired from Soviet submarines in the Eastern Baltic. In the same period came the evidence that Beriev was designing modern jet-engined flying-boats for the expanded Soviet naval air arm.

To-day most estimates give the Soviet Naval Air Force an operational strength of some 3,500 aircraft with a personnel strength of 80-100,000 men including both aircrews and ground staff. This compares with an estimated strength of some 3,000 aircraft in the middle 1950s. Of the present force some 1,500-2,000 aircraft are jet fighters, subsonic and supersonic, mainly MiG 17 and MiG 19 fighters. The former have a top speed of some 725 m.p.h. and the latter are capable of reaching a speed of about Mach 1·4. In addition a number of MiG 21 supersonic fighters have come into service in the last year or so capable of flying in excess of Mach 2, that is more than twice the speed of sound. The main bomber force consists of about 750-1,000 twin-engined bombers of which the bulk are Ilyushin 28 and Tupolev 14 machines. The Tu 14 is the first Soviet jet land-based plane to be specifically and entirely assigned to the naval air arm. Dubbed the *Bosum* by NATO, it is a twin-jet machine equipped with Klimov turbo-jets each with a thrust of some 6,000 lb. It has an aircrew of either 3 or 4 and a radius of action of 800-1,000 miles, i.e. a range of over 1,800 miles which can be increased to over 2,500 miles with the extra tanks which are available. It can carry torpedoes, mines or conventional or atomic bombs. It would probably be limited initially to operations in the Mediterranean, the Black Sea, the Baltic and North Seas and the Arctic. It is not a strategic long-range naval co-operation aircraft and is in roughly the same category as the British Canberra.

The most significant change in Soviet naval air equipment in the middle 1950s was the creation of a force of long-range four-engined strike and reconnaissance aircraft which were clearly intended to play the same rôle in anti-shipping and anti-submarine warfare, that was allotted to the German Focke-Wulf 200 squadrons which had carried out such successful naval air operations from Bordeaux and various Norwegian bases in the Second World War. At first, Kuznetsov's long-range squadrons had to be content with piston-engined Tupolev four-engined machines which were becoming obsolescent in the Soviet Air Force Strategic Air Command (*Dalnaia Aviatsia*) in 1954-55 and which were transferred to Soviet naval air regiments from the Strategic Bomber Command of Marshal Sudetz. But this was a temporary equipment phase. Since 1956 the Soviet naval air forces have been equipped with both Tupolev 16 twin-jet bomber planes and the longer-ranged and more powerful Myasischev Bison four-jet bombers. At the moment there are perhaps between 100 and 200 of these Soviet jet bombers in Soviet naval air regiments. Clearly the situation is likely to change crucially in the next few years. As Soviet land-based intermediate-ranged rockets and intercontinental ballistic missiles increase in total and in accuracy, the rôle of the Soviet DA strategic land-based bomber arm, with its Tupolev, Ilyushin and Myasischev designed subsonic and supersonic jet bombers will become much less important. In addition, as the number of atomic-engined ocean-going U.S. submarines capable of firing Polaris long-range atomic missiles increases, so the Soviet anti-submarine naval and air forces must surely also be increased. It would therefore be logical for the Soviet Defence Ministry to divert a high proportion of the present Soviet strategic bomber force to the tasks of anti-submarine and anti-shipping warfare such as long-range ocean reconnaissance, co-operation with Soviet submarines in attacks on convoys and enemy submarines and finally in long-range air attacks on convoys and submarines using torpedoes, guided bombs and other guided air-to-surface weapons.

We have so far learnt very little about Soviet air developments in this direction. The news was first of the development of high-speed jet fighters in the late 1940s and early 1950s, of medium and heavy jet bombers in the middle 1950s and of a series of long-range jet transport planes in the late 1950s. But the accelerated developments in Soviet long-range rockets in the late 1950s must surely portend a release of strong forces of land-based jet bombers for other purposes. The most logical use of them would be to strengthen the long-range strike and reconnaissance arm of the naval air forces of the USSR. Potentially there is no reason why this bomber reconnaissance naval air force should not be raised in strength to 1,000 long-range jet-engined machines in the 1960s. This would compare with a force of less than 50 Focke-Wolf 200 aircraft which was all the Luftwaffe operated in World War II in a similar anti-shipping rôle. In view of the emergence of Soviet airborne radar both for detecting ships and guided bombs and missiles and of the increasing range and speeds of Soviet long-range bomber aircraft—the latest delta-wing bomber to come into service has an intercontinental radius of action and supersonic top speeds—it is clear that the menace of the Luftwaffe four-engined Focke-Wulf 200s to shipping was infinitely small compared with the present and future potential of the Soviet naval arm. And yet the Focke-Wulf squadrons played no mean part in the naval air battle of the Atlantic especially in 1940-41.

Another Soviet naval air development of the 1950s has been the use of helicopters for patrol and reconnaissance from cruisers, such as the Alexander Suvorov and the Sverdlov, and for land-based air-sea rescue, anti-submarine and patrol reconnaissance operations. In addition Soviet helicopters designed by Kamov are being used from Arctic naval air bases for ice patrols. The current Kamov helicopters in use in Soviet naval air squadrons are mainly the two-seater KA 15 and the four-seater KA 18, but improvements in performance and load capacity are to be expected in the next few years, for the USSR has shown in the 1950s that both in military and civil fields she is probably more conscious of the all-round importance of helicopters than Western air powers. Certainly since the 1956-57 period there has been intensive anti-submarine operational training in the Soviet naval air arm using Kamov and MIL helicopters. The MIL 4 helicopter, with a useful load of 2 to 3 tons, and powered with a 1,700 h.p. radial air-cooled engine is used chiefly for anti-submarine patrols while the smaller MIL 1 and 3 are used for air-sea rescue and patrol work. The latest versions carry improved ultra-short-wave radio sets. It would however be surprising if the larger land-based helicopters, such as the YAK 24 and the MIL 6, capable of carrying greater loads at higher speeds and for longer distances are not introduced into the Soviet naval air arm in the near future.

At one stage it seemed as though the Soviet naval air arm was going to develop a large force of long-range anti-shipping flying-boats designed by Beriev. Back in 1932 Beriev had built for the Soviet naval air forces his first short-ranged reconnaissance flying-boat fitted with a 650 h.p. M 17 engine. It was called the MBR 2 (the initials stand for naval bomber and reconnaissance aircraft) and it was followed by a more powerfully engined version in the later 1930s which was called the Beriev 2. A few are still engaged on Soviet ice and fishing patrols. Just before the outbreak of the Germano-Soviet War Beriev designed two new naval aircraft which became standard equipment in the ship or water-based naval air forces of the USSR. These were the two-seater biplane KOR 2 single-engined floatplane reconnaissance machine, which had a later version called the KOR 2 with a rather more powerful engine of 1,100 h.p. (the M63 radial); then he designed the long-range reconnaissance flying-boat the MDR 6 or BE 4. It had a maximum speed of about 250 m.p.h., a crew of either five or six and a defensive armament of two 7·62 mm. machine guns. But neither of these two naval aircraft played any important rôle in the Second World War, partly because they were produced in very small numbers, but mainly because it was not Soviet air policy to build up a force of floatplanes and flying-boats and use them in any key reconnaissance, bombing, mine-laying or torpedo-dropping rôles.

In the immediate post-war period Soviet air policy remained the same; no new floatplanes are known to have been produced and only one or two flying-boats. The best known and most used is the BERIEV 6, a twin-engined flying-boat which first came into some general use in Soviet naval air squadrons in the early 1950s. Called the *Madge* by the NATO christeners of Soviet aircraft, it carries a crew of 8, has defensive 23 mm. cannon armament in the nose, dorsal and tail positions, a top speed of about 260 m.p.h. and a maximum range of about 3,000 miles. It is likely to be used mainly for local reconnaissance and patrol work in the Black and Baltic Seas, and the Pacific and Arctic Oceans. It can carry a load of about 2-3 tons of mines, torpedoes, bombs or depth charges. It has a wing-span of some 108 feet and is likely to be highly vulnerable in combat to modern jet fighters either land-based or carrier borne. One would have imagined that the Soviet naval air authorities would have asked for more modern jet-engined flying-boats or would have decided not to use flying-boats at all for operational duties. A few years ago it was reliably reported that Beriev was designing both twin-and four-jet engined flying-boats with better performances, but so far no Soviet naval air unit is known to be equipped with any of them. And now that the naval air force is being refitted with land-based four-jet Bison and other long-range bombers it would seem unnecessary for these squadrons to be fitted with long-range jet-engined flying-boats.

Whatever current assessment one makes of the operational importance of the Soviet naval air arm is likely to be wildly obsolescent in the next few years for two reasons, one of which has been already indicated. This is the potential increase in land-based four-jet bombers in both the subsonic and supersonic categories which could result from the further growth and accuracy of Soviet intercontinental ballistic missiles which would replace to some extent the present strategic bomber forces of the DA long-range air command. But the emergence of a strong force of Soviet ocean-going submarines capable of launching long-range atomic missiles would certainly boost the significance of the Soviet Navy and naval air forces. It would create the need to have a stronger and more modern naval air force capable of co-operating to the full with the new long-range strike atomic weapon-carrying submarines. One Soviet writer, Shavtsov, a senior naval captain, put the new position this way, in an article written in 1955 in *Military Thought*: "The employment of the atomic weapon in combat at sea makes it necessary to raise operational requirements especially for reconnaissance and air defence." And so the future development of Soviet long-range missiles is closely connected with the future of the Soviet naval air arm. It has been said by many strategists in the past few years that the long-range ocean-going submarine, capable of carrying and firing a dozen or more nuclear and even thermo-nuclear rockets will probably create a revolution in the pattern of defence policies in the 1960s and thereafter.

That the Soviet Union is developing long-range missiles to be fired from submarines has been known for more than a decade. Firing trials from submarines in the Baltic using a version of the German V 2 were reported by eyewitnesses as early as 1950 as has been stated elsewhere in the book. This was not surprising, for in 1945 the Soviet Navy acquired from the spoils of Peenemunde a special German rocket-launching carrier and German equipment for towing three V 2 rockets behind their submarines. During the last ten years the USSR has been actually developing two classes of submarine-launched rockets, the Komet and the Golem, though these rockets may also be used for land-based and even air launchings as well. The Komet 1 was the first of the series of the submarine-launched rockets. It is variously reported to have a range of between 100 and 200 miles, and a number of military experts have stated that it has been in large-scale production for several years. In the view of some naval specialists a force of Soviet submarines is already fitted with this long-range missile which carries a nuclear war-head. But it is the successor to the Komet 1, the Komet 2, which presents a major threat to most inland cities of the United States and other territories. The Komet 2 is thought to have a range of about 600 miles. Rear-Admiral H. G. Rickover, a pioneer in the development of nuclear propulsion for U.S. submarines, in an official statement before a U.S. Senate Committee made in January 1958, expressed the view that Soviet submarines were then already fitted with missiles having a range of more than 200 miles and capable of carrying nuclear warheads. He may have been referring to the Komet 2 or possibly to the Golem 2 reported to be a two-stage liquid-fuel rocket with a range of over 1,000 miles. Other reports during the last two or three years have indicated that both are in service in the Soviet submarine arm. The Komet 2 and perhaps the Golem 2 still could not reach all targets in the United States; but, as Admiral Burke of the U.S. Navy has pointed out, over half the major cities of the United States are within 100 miles of the coast. However, during the last year or so the development of the Komet 3, the Soviet equivalent of the solid-fuel U.S. Polaris, with a range of 1,500 miles has enabled the Soviet submarine arm to be potentially the most powerful strategic air weapon in Soviet hands. It may of course take a decade or more for this to happen. At the moment both Soviet and Western submarines in the ocean-going category could be immune from anti-submarine detection devices. No one can predict how many of the 500 submarines in the Soviet Navy will have crews trained to the high standards which ocean-going underwater warfare demands.

In any event only just over half of them probably have the range to engage in strategic naval warfare, and no doubt only a percentage of these will be permanently in commission and ready to strike accurately. The mapping of the ocean-beds, the emergence of Soviet nuclear-powered submarines (there are probably less than half-a-dozen in commission at the moment), the standards of Soviet long-distance communications, of Soviet strategic naval air co-operation, of Western anti-submarine strength are but a few of the factors which will influence the position and potential of the future long-range air striking power of Soviet submarines. But there is clearly a possible major change in Soviet defence strategy looming ahead in the 1960s. There is no reason why Soviet destroyers and some cruisers should not be fitted with long-range missiles. Indeed there are reports that some of them are already fitted with them. The new strategic striking power of Soviet submarines may call for some reorganisation of the Soviet Defence Staffs producing perhaps a combined navy, army and air long-range strike force combining the assets of the DA strategic bombers, the land-based long-range rockets and the submarine long-range strike force.

To sum up, the Soviet naval air arm and naval long-range rocket striking power may well become the most important strategic arm of the Soviet armed forces. Because the USSR was late in using both its ships and naval air forces in any important strategic rôle, this potential did not emerge until the 1950s, which saw a modernised Soviet naval air arm, fitted with subsonic and supersonic jet fighters and bombers; it now has a useful force of anti-submarine helicopters and a long-range bomber reconnaissance land-based air component, capable of co-operating with ocean-going submarines. But for the reasons given in this chapter, the shape of naval and naval air affairs in the USSR is likely to be greatly different in the next ten years. One wonders if the Soviet Army, with its special pride of place in the Kremlin, will not endeavour to exercise some major control over future Naval Air Force developments. It is surely significant that Soviet Army ranks have been retained throughout. But certainly the Soviet Naval Air Force, for so long a relative Cinderella in the Soviet military hierarchy, is likely to become more and more a major element in an age in which submarine warfare may become more than ever the key to long-term offensive strategy.

CHAPTER 10

TACTICAL AIR POWER

IN the Second World War the tactical air forces of the Soviet VVS comprised two-thirds or more of the front-line operational strength of the entire Soviet Air Force. In addition both the naval air forces and the long-range ADD air regiments were also extensively used in tactical rôles in support of the Red Army land battles. It is probably true that Soviet air power concentrated more on the direct support of ground forces than any other air force of the 1941-45 period. Some sources indicated that at times ninety per cent of the Soviet Air Force operational strength was so engaged.

Despite the growth of Soviet strategic air power for both offence and defence in the past ten years, the tactical air forces have remained by far the strongest arm numerically and the advent of tactical atomic rockets has not yet modified this general position to any extent. Although the equipment of the Soviet tactical air forces has been modernised and improved continuously throughout the post-war period, the underlying principles on which it would operate in the future do not differ greatly from those which governed its operation in the last year or so of the Second World War. In fact it is now much more capable of putting its principles into practice because of the improvement in the mobility of its aircraft squadrons and equipment and in the standard of its radio communications.

These principles were restated by Colonel-General Sudetz when Chief of Staff of the Soviet Air Force immediately after the war, and it seems that they still apply despite the advent of tactical atomic weapons. The first and most important principle is the establishment of air superiority by the destruction of enemy aviation in the air and on the ground. To achieve this end the Soviet tactical air forces and strategic air forces have a combined front-line strength of over 10,000 jet fighters. It is in the field of single-seater day fighters that the Soviet Air Force first achieved virtual technical parity with the West when the MIG 15 fighter went into production in 1947 and began to be delivered to Soviet Air Force units in 1948. But although this aircraft showed new high standards of workmanship for Soviet fighter planes and was well equipped and armed by Western standards, the lack of a radar gunsight proved a severe handicap in the Korean War operations. Clearly the Soviet tactical air force needed new weapons and equipment to ensure the achievement of local air superiority by its fighter aircraft.

The diversion of the MIG 15 to a fighter-bomber rôle in the 1950s and the designing of the MIG 17 and MIG 19 jet fighters to keep pace with U.S. fighter aircraft development had a dual purpose: to strengthen the strategic air defences of the USSR and to increase the possibility of securing and maintaining local air superiority over battle zones. As several U.S. Sabres were captured by the Communists in the Korean War it is conceivable that the radar gunsights from these aircraft served as a basis for the Soviet model which was fitted as standard equipment on the MIG 19, the first Soviet jet fighter capable of supersonic speeds in level flight. In addition to greater accuracy through the new radar gunsight, fire power was raised by stepping up armament from two 23 mm. and one 37 mm. cannon, by adding a further 37 mm. cannon and unguided air-to-air rocket missiles of 50 mm. calibre, of which 32 could be fired by the later MIG 19 models. The MIG 19 fighter was powered by two Lyulka axial-flow turbo-jet engines each providing nearly 9,000 lb. static thrust with the use of an after-burner. By 1957-58 it was gradually being replaced by the MIG 21 fighter powered by a single turbo-jet engine giving a maximum thrust of some 17,000 lb. and a top speed of over Mach 2. In the last year or so an even faster and more powerful jet fighter the MIG 31 has been coming into service. This may be the delta-wing fighter for which the Soviet authorities claimed a world speed record in the autumn of 1959. In addition to developing faster and more powerful day fighters, two types of guided air-to-air missiles have been coming into squadron service in the past few years, of which one has infra-red guidance mechanism.

It is now clear that statements by Soviet military and political leaders made in the autumn of 1957 that the manned fighter was to become a museum piece have no substance in terms of the current trend of Soviet weapons development. This is not surprising. In the age of atomic weapons ground targets in ground battles are likely to be even more mobile and dispersed and so more subject to attack by manned aircraft rather than by rocket-guided weapons, which usually need a fixed point to aim from and a fixed point to aim at if they are to be accurate. And if rocket weapons are to use present forms of guidance, such as infra-red detection or beam navigational aids, in the *va et vient* of the ground battle, they are more likely to hit their moving target if fired from light or medium bomber aircraft.

Certainly the development of Soviet atomic artillery and tactical atomic rocket weapons, such as the T 1 rocket, firing distances of 250-500 miles, have not limited the rate of development of conventional tactical air weapons. In parallel with the continued technical improvement in the single-seater fighter force there has been progress in the development of all-weather fighters and medium-range tactical bombers and fighter bombers. The rôle of these tactical air forces seems to be potentially the same as when they were used in World War II, when their missions in support of the Soviet ground forces were to attack and destroy enemy troops and equipment on the field of battle, to attack and neutralise enemy reserves of men and equipment, to limit supplies to enemy forces in the battle area, to combine in action with the ground forces during offensive action and finally to speed up by bombing the process of the pursuit of the enemy.

As far as one can see, the Sovict are developing both conventional and atomic tactical weapons in parallel with two things in mind. First they would initially use conventional bombs, rockets and guns in a local tactical battle in order not to induce the West to raise the stakes of the local fight by the introduction of tactical atomic weapons. But even more important is the uncertainty of the effect of tactical atomic weapons on the course of the ground battle. It could well be that the best initial Soviet reply to the potential use by NATO of such atomic weapons as the Corporal, Sergeant, Honest John and Little John is to apply such overwhelming local conventional tactical air power that tactical atomic weapons could not be used to full effect and the need to bring Soviet atomic weapons into play would not be so compelling. If for instance Soviet airborne troops used tactics similar to those employed in World War II, when parachutists jumped close to their objectives and went immediately into action, the use of atomic weapons against such forces would be difficult if not impossible. And if Soviet ground troops carried out their initial attacks at night and followed up with use of airborne troops among Western troop dispositions, it is hard to see how atomic ground-to-ground weapons could be used without great danger to friendly ground forces. Atomic weapons need worthwhile targets which may take a few hours to find. Even in their local tactical rôle they are not likely to be used immediately and indiscriminately. In that key period of a few hours, a strong local concentration of conventional fighters and fighter-bombers could do much to neutralise attempts to reconnoitre worthwhile targets and later the means of firing at them accurately with atomic weapons.

In World War II concentrated local air attacks were mounted by Soviet Air Force tactical units within half an hour of the call for the air strike. The target was usually 50-150 miles away. To-day there is no reason why with better radio communications and higher speeds an air strike of major dimensions could not be mounted within less than half an hour against a target which is 300-400 miles away.

There is a tendency in some circles to underrate the potential of Soviet tactical air power. In a report *The British Army in the Nuclear Age*, published in November 1959 under the auspices of the British Army League, it was stated (page 21): "There is a good reason to suppose that the Western air forces would be a good match for the Russians who have not so far especially distinguished themselves in the air. It is for instance unknown whether Russia's airmen and machines could bomb as accurately and effectively as those of the West, in spite of the brilliant performance of a number of the Soviet Union's leading aircraft designers; certainly Russian air power was a disappointing factor in the Second World War." Any thinking which goes back to Soviet tactical air operations of the 1941-45 period is surely likely to be misleading. There have been many fundamental improvements since which would make themselves felt. The standards and equipment of Soviet aircraft have improved out of all recognition. Whereas they fought some of the air campaigns of the Second World War with fighters of inferior performance, there is nothing to suggest that present-day Soviet fighters are inferior to those in NATO air forces. Whereas their signals and ground-to-air communications were greatly inferior to those in Western air forces in 1941-45 the present-day gap is now probably negligible in this respect. There is evidence that Soviet training in escorting tactical bombers has reached high standards and it is difficult to see why

Soviet airmen cannot bomb and strafe ground targets in a battle area as effectively as NATO airmen.

But the great advantage that Soviet tactical air forces are likely to have in any future European struggle would be their greater ability to concentrate large forces of fighters and light bombers at the *Schwerpunkt* of the ground battle. It has been emphasised again and again that NATO ground divisions are a defensive shield force. It is thought that the twenty to thirty divisions of the Western Alliance might hold up an opposing force of three times that size and that even then the Soviet task might be made difficult unless they could secure a reasonably strong measure of air superiority. But once it is accepted that NATO forces are there initially for defence and not attack, the Soviet problem of establishing local air superiority becomes much easier. It is unlikely for instance that NATO tactical air forces from southern Europe would be thrown into an air battle over Germany, nor could the tactical air forces of NATO in Turkey be used in a local battle for air superiority over Germany. Since NATO ground forces constitute a defensive shield or trip wire, it must be assumed that the time and place of initial ground thrust by armoured forces or airborne troops would be left to the Russians to decide. In such circumstances they can concentrate strong fighter cover and tactical air forces in the sectors that they choose; moreover it is hard to see how effective interception of high-speed Soviet jet light bombers can be organised by NATO fighters, and it will surely be unlikely that ground-to-air guided missiles however prolific would be generally effective against low-flying Soviet tactical air squadrons.

Apart from the important advantage of local initiative, the Soviet tactical air forces are likely to have a great numerical preponderance, and this is in itself an important element in the application of tactical support. Whereas strategic fighter defence and strategic bomber operations call for the highest standards of planning, execution, personnel training and aircraft and aircraft equipment, tactical air operations do not rely on these basic factors to the same extent. It can be assumed that Soviet pilots would not lack courage and determination nor would their aircraft be technically inferior in performance. If they can maintain a series of dawn-to-dusk attacks in the early stages, they might be able to limit ground movement among NATO forces to a considerable extent. There is a tendency to regard

NATO's tactical atomic weapons, i.e. the ground artillery, rockets and air-launched bombs as a stabilising deterrent to meet the Soviet preponderance in conventional ground forces; but quite apart from current Soviet growth in atomic artillery and other tactical atomic weapons, the further development of conventional Soviet tactical air power can negative to a considerable extent the military value of NATO's tactical atomic weapons. NATO tactical air forces must obviously achieve some measure of local air superiority or at any rate parity if their atomic weapons are to be effectively used. The Soviet Air Force can develop the capability of immobilising ground movement and supply operations in Western Europe to the extent that Anglo-American tactical air power did in the summer of 1944 or indeed the operations of the Luftwaffe in the summer of 1940.

It is clear that the Soviet High Command regards conventional tactical air power as having as high an importance in the age of nuclear weapons as it had during the Second World War. It has added extra stress on mobility by the provision of a growing air transport command, whose strength has been conservatively estimated at 2,000 aircraft and, too liberally perhaps, at as many as 5,000 aircraft. The difference between these two figures has a possible explanation. The lower figure of 2,000 aircraft may include nothing but the declining force of twin-engined transport Ilyushin 12 and 14 aircraft and the growing number of Antonov twin- and four-engined turbo-prop. aircraft and of Ilyushin four-engined turbo-prop. machines together with the larger type transport helicopters such as the YAK 24 and MIL 6. The higher figure of 5,000 probably includes the Aeroflot civil passenger planes and Antonov 2 biplanes used for parachute training or agricultural work under Aeroflot aegis. But even taking the lower figure of 2,000 transport machines it is reasonable to assume that Soviet military transport regiments can provide a two-divisional lift of parachute or airborne troops or carry an equivalent amount of equipment and supplies. It is also a reasonable assumption that current air lift potential will be doubled at least in the next five years as the piston twin-engined Ilyushin transports are replaced by more powerful, and longer-ranged, Antonov and Ilyushin turbo-prop. aircraft.

One of the most difficult assessments is to predict the future value of Soviet helicopters in any future ground fighting. Assuming that a high degree of local air superiority can be achieved, helicopters such as the Yak 24 and Mil 6 will be invaluable to fly troops, artillery, ammunition, staff cars and trucks to the battle area. For five years now the USSR has been using large helicopters in these rôles in annual military manœuvres. The following up of the Yak 24 which can carry 40 men, by the Mil 6 which can carry twice that number, indicates continuing Soviet faith in the ability of helicopters to achieve mobility in the battle area. And yet these slow vulnerable transport machines can surely only be operated under a very strong air umbrella of fighters if they are being used to support a local ground offensive. And so the importance which the Soviet armed forces attribute to using large numbers of troop-, weapon- and vehicle-carrying helicopters is another reason for believing that the USSR will continue to strengthen her conventional tactical fighter air power as far as overall resources permit. If ever Soviet ground forces were committed to combat, with or without the use of tactical atomic weapons, one can anticipate an old-fashioned tactical air blitz on enemy airfields, intense dog-fights with enemy fighters and strong fighter patrols, in other words the conventional World War II pattern for establishing local air superiority. Once this is done, and it seems that the Soviet air forces might achieve this fairly quickly, then transport aircraft and helicopters will be extensively used. To throw these helicopters into battle before local air superiority has been achieved would be courting the same disastrous losses as the Soviet Air Force suffered in their unescorted or ill-escorted days of 1941-42. The local air superiority they achieved in the Stalingrad battles at the end of 1942 and early in 1943 are also a reminder of how their air transport supply and reinforcement operations succeeded in these conditions while the German effort failed.

There is however one aspect of Soviet tactical air power which may cause the USSR to hesitate to plunge into war in Western Europe. This is the uncertainty of its communications in Central Europe. It must be as hard in Washington as it is in Moscow to assess their vulnerability to sabotage and satellite disaffection. There are long lines of communication running through East Germany, Poland and Hungary and the advanced air bases of Soviet tactical air forces lie well inside these territories. The uncertain elements in Hungary and Poland were demonstrated in 1956 and there must surely be some doubts in the Kremlin as to what would happen if East German air bases had to be used for attacks on West Germany. It may well be that the uncertainty of tenure of Soviet air bases in Germany, Poland and Hungary is acting as part of the deterrent to war. It is a deterrent which could be as important as the stock piling on each side of nuclear weapons. The Soviet Air Force must have secure bases from which to exercise air superiority and it may well be thought in Moscow that the present advanced bases might be a liability rather than an asset.

However, in purely military terms it is the use of tactical atomic weapons which forms the main deterrent to Soviet ground forces and the likelihood that their use would provoke the escalator element, i.e. the resort to the main deterrent consisting of megaton thermonuclear weapons. Since it is reasonable to assume that the Kremlin is as anxious as the West to avoid the escalator effect which eventually puts to the supreme test the present balance of East-West annihilation potential, how can the USSR avoid using atomic weapons of any kind, even those of small yield in a tactical rôle? The first method would be to act as in Korea, that is not to commit any major military units, but to leave the initial fighting to her allies and to provide them with the technical air help and equipment that were needed. The present re-equipment of the East German Luftstreitkräfte with supersonic MIG fighters would be a typical instance. If then the situation deteriorated from the Kremlin's point of view, they would probably avoid throwing in any concentrated major assault forces which would provide obvious targets for Western atomic weapons, and they would concentrate on restoring the position on Communist held territory. At this stage they would probably throw in their maximum available conventional tactical air forces and would be unlikely to use any tactical atomic weapons. In order to avoid the fatal stepping up of the stakes they must surely try to rely on conventional air power and ground forces until the very last moment; and yet even to make the major use of Soviet ground forces and conventional tactical air power might be to court disaster. This would probably involve the penetration of NATO territory and so the use of the Western shield forces. Since these are likely to be ineffective if confined to the use of conventional explosives the Soviet must surely appreciate that their conventional ground and tactical air forces are likely to be quickly opposed by Western tactical atomic weapons. Once these are used effectively, the entire pattern of conventional ground fighting will be disrupted. If the results of Operation Carteblanche, the NATO tactical atomic air manœuvres of 1955, were even the vaguest of foretastes of what is likely to happen, communications and airfield serviceability will immediately become chaotic and the consecutive direction of operations in the battle zone will become virtually impossible.

In order to avoid this situation and to avoid escalation into the realm of the use of strategic nuclear weapons, it is my belief that Soviet conventional tactical air power will continue to be strong and further modernised in the hope that it can largely neutralise NATO tactical atomic weapons before they can be brought into use on a large scale. The Kremlin probably appreciates that some NATO weapons with atomic warheads will not be in the possession of front-line ground and air forces but will be kept back at divisional or higher level pending a political decision to use them. This is also believed to be the position of atomic weapons in the Soviet armed forces too, with the Communist Party exercising control via a special fourth arm of the defence forces controlling all atomic weapons. Thus, so the Kremlin may argue, there might be an initial period of a few hours during which they might be able to establish local air superiority of such a nature that the decision to use atomic weapons by NATO leaders would be largely ineffective.

It is this which constitutes the great importance of the conventional tactical air forces of the USSR and which may explain why large and modern air-ground support forces of over 10,000 aircraft have been retained in the 1950s, in a period in which one might have imagined that Soviet atomic artillery and tactical guided missiles would have largely replaced the tactical bomber forces. But the fact is that throughout the post-war period there has been continual modernisation and improvement of conventional tactical air power in the USSR and the advent of Soviet tactical atomic weapons in the last five years has not yet halted or slowed up this process.

Until the jet age of the 1950s, the USSR tactical air forces continued to use conventional engined tactical support aircraft such as the Tu 2 and the Ilyushin 10. That is to say the first post-war priority was to convert the fighter squadrons to jet-powered machines such as the Yak 15 and Mig 15 and to leave the tactical bomber regiments with their piston-engined equipment for the time being. Even so there were armament and equipment improvements made in the late 1940s. The 23 mm. cannon were replaced by the heavier

37 mm. cannon on the IL 10. There was more general equipment of the Ilyushin Sturmovik regiments with rocket projectiles of which eight could be fired from each aircraft, each weighing fifty-six pounds; and the Soviet RSB transmitters for maintaining wireless contact with ground forces were improved and more generally installed in individual Sturmovik and fighter-bomber aircraft. In another well-tried tactical bomber, the PE 2, the Soviet tried out their first ejector seat in the late 1940s and these were also to become standard equipment during the 1950s.

It was not that jet-engined ground-attack aircraft were neglected in the 1945-50 period but rather that none of those which emerged ever went into large-scale production to become standard equipment. The twin-jet MIG 9 which was in unit service at the end of 1947 was powered by German designed axial turbo-jets which were not powerful enough to meet Soviet requirements. The alternative armament with 53 mm. and 37 mm. cannon indicated the experimental period through which Soviet tactical air equipment was passing. Another light tactical jet bomber which emerged in this pre-Korean War period was designed by Andrei Tupolev, the doyen of Soviet aircraft designers. This machine was also powered by German designed axial jet engines but much of the wing, fuselage and tail designs were taken from Tupolev's successful piston-engined TU 2 bomber. Tupolev's twin-jet attack bomber was an attempt to produce a Soviet equivalent of Britain's emerging Canberra bomber which then led Western designs in this category. But Tupolev's attempt to produce the first standard Soviet twin-jet bomber for Russia's tactical air forces was not successful; from 1950 onwards he was to concentrate on the design of long-range jet bomber and transport aircraft and to leave the design of the new Soviet jet tactical aircraft to teams of designers working under Ilyushin and Yakovlev.

It was in the early days of the Korean War, that is the summer of 1950, that the first reports came of the Ilyushin 28 twin-jet tactical bomber re-equipping Soviet air regiments stationed in East Germany and Siberia. It is likely, however, that re-equipment began rather earlier, as the test-flying of the IL 28 had taken place in 1948 and over 50 of these aircraft took part in the May Day 1950 fly-past over Moscow's Red Square. The IL 28 was powered by Klimov turbo-jets developed from the Rolls-Royce Nenes which the USSR had purchased from England. With a maximum speed of some 580 m.p.h., a maximum range of some 2,000 miles and a maximum bomb load of some 6,000 lb., this Soviet machine was to become the most widely used in Soviet tactical air regiments in the next five to ten years. It is likely that more than 10,000 of them have been built so far, for they have not only equipped the tactical air units of the Soviet Air Force but also those of the Czech, Polish, Roumanian, Hungarian, East German, Chinese, Afghan and United Arab Republic Air Forces.

The combination of the IL 28 tactical bomber and the MIG 15 fighter made it clear that the technical gap in 1950-51 between East and West in the field of air-ground support was narrowing. But it still needed better Soviet radar equipment and more mobility if the gap were to be finally closed as well as setting a higher standard of combat training and pilot selection. In addition the design and performance of Soviet aircraft had to keep pace with the West in the new supersonic jet age of the middle 1950s. As far as one can judge, the USSR has succeeded in doing this. By 1953-54 the MIG 17 was replacing the MIG 15 in both the strategic defence and tactical air units of the Soviet Air Force. The new fighter was later equipped with airborne radar which made it a more accurate gun-firing platform, extra tanks were added and also an efficient after-burner to its Klimov jet engine which brought its top speed to over 700 m.p.h. And in 1955 the MIG 19 supersonic jet fighter began to replace the MIG 17 in tactical air units, revealing that the USSR had a counterpart to the U.S. F 100 Super Sabre. And within a year or so there were further examples of technical progress to report. A new and more powerful fighter, the MIG 21 fitted with a turbo-jet engine giving more than 15,000 lb. static thrust and an auxiliary rocket engine, began to come into service; it had a top speed of over 1,000 m.p.h. and perhaps as much as 1,200 m.p.h., equivalent to about Mach 1·8. In the same period (i.e. 1956-57) a supersonic improvement on the Ilyushin 28, called by NATO the "Blowlamp" and by the Soviet Air Force, the Ilyushin 140, began to come into service in the Soviet tactical air units or the "Front Aviation" as they were now called. There can be little doubt that during the next two or three years Soviet tactical air forces will gradually become almost entirely supersonic in both the fighter and light bomber air regiments.

The standard night and all-weather fighter in the Soviet Air Force is the YAK 25. It is fitted both with AI radar equipment for ground-controlled interception and with radar sights for firing its cannon.

It began to come into regular service as a fighter in 1954-55 and a year or so later a bomber version of it with radar bombing equipment began to come into use in Soviet tactical air regiments ensuring them a day and night and all-weather capability in combat. Within the last two or three years a supersonic version of the YAK 25 twin-jet has been equipping both the night-fighter and tactical bomber units of the Soviet Air Force. And as the manned fighter is gradually phased out of Soviet strategic air defence in the next two or three years, larger numbers of supersonic Yakovlev twin-jet machines are likely to be available for Soviet tactical air forces. And as ground-to-air guided missiles such as the T 6 and T 8 come into more general service in the PVO air defence units, a higher proportion of advanced day fighters such as the MIG 19, MIG 21 and its Mach 2-3 successor, the MIG 31, as well as the Sukhoi delta-wing fighters will be released to secure local air superiority over the battle area. Thus the present force of 5,000 tactical subsonic and supersonic jet bombers, which include a force of MIG 15 and MIG 17 fighter-bombers, in addition to the aircraft types already mentioned are likely to have a fighter force of up to twice this size available for escort patrol and interception work over the battlefield.

Current technical developments in the conventional tactical air forces of the USSR suggest that the USSR would be disinclined to use tactical atomic weapons unless compelled to. Russia is reported to have a growing force of atomic artillery and mortars with a range of 15-20 miles as well as large numbers of tactical atomic guided-rocket weapons with ranges of from 100 to 500 miles, such as the T 1, T 5 and T 7. In addition she has a large supply of small-yield atomic bombs which could be dropped by her conventional tactical air regiments. But her decision to keep all her atomic weapons under one central military-political authority and not to distribute any tactical atomic weapons, either artillery, mortars, rockets or bombs, to any of the Warsaw Treaty powers reinforces the notion that the Soviet Union will do all it can to avoid using them. Indeed the Kremlin is in a much better position than NATO to avoid the use of tactical atomic weapons since, as far as is known, the allies of the USSR not only do not have them but, as far as is known, show no signs of wishing to acquire them. China may be the first exception to this rule. In this context it is of passing interest that whereas Molotov indicated that the USSR was providing China with atomic bombs in the middle 1950s, Khrushchev has not confirmed that this has happened nor that it is likely to. Nor is there any indication from any other sources that this has in fact yet happened.

The aspect of Soviet tactical air equipment about which surprisingly little has been published is that of air reconnaissance. It has been reported that the legacy of German air cameras has been fully exploited and that there are reconnaissance versions of the latest Soviet single- and twin-engined jet fighters and fighter-bombers, and that special tactical air regiments have been set up which are more specialist in training, equipment and in manœuvres than the reconnaissance units of the Second World War. Then there was a tendency for tactical air reconnaissance to be carried out by the specially skilled pilots of the fighter and fighter-bomber regiments and for results of visual observation to be reported orally. But in the post-war period a tradition of reconnaissance specialisation has grown up, with aerial photography, radar photography and lately the experimental use of television becoming more general. It is of course virtually impossible to say how efficient and effectual the new equipment would be under combat conditions. Certainly the likelihood that NATO will be compelled to use tactical atomic weapons if war breaks out in the next few years, imposes on the air forces of the USSR the need for more detailed tactical reconnaissance coverage than ever before. It would be surprising if some of the new Kamov, Mil and Yakovlev helicopters were not brought into use at any rate as artillery spotters for the short-range atomic artillery and mortars. But believing that one of the USSR's first objectives in a local war would be to negative the effect of Western tactical atomic weapons, it would seem that they can best achieve this by intensive reconnaissance of air and ground movements combined with airborne and parachute operations to disrupt rail and road communications and intensive fighter sweeps to deny air transport flights, accompanied by intensive conventional tactical bombing of the NATO battery and rocket sites in the battle areas. With a force of over 100,000 airborne troops, over 10,000 tactical support aircraft and helicopters available, it may be unnecessary for Soviet armed forces to concentrate in the early days of combat a large number of divisions which would provide a tempting target for NATO atomic weapons. There is still much debate whether tactical atomic weapons favour the attacking or the defending forces. There can however be little debate as to whether strong tactical conventional air power with local air superiority can negative the effectiveness of tactical atomic weapons to a considerable extent. I believe that Soviet tactical air power will increasingly be able to achieve this in the next year or so and enable the Kremlin to take a temporary decision not to use tactical atomic weapons automatically if the West should find itself compelled to do so.

CHAPTER 11

MILITARY AND CIVIL AIR TRANSPORT

IT may seem strange to fuse an account of Soviet military and civil aircraft transport resources into a single chapter; but the fact is that civil aviation is more closely linked to the military hierarchy in the USSR than in any other country in the world. At the moment the expanding elements of Soviet civil aviation constitute a separate administration under the Council of Ministers, independent of the Soviet Defence Ministry and the Soviet Air Force. But during the Germano-Soviet War and up to 1948 civil aviation was subordinate to the Defence Ministry and fully merged into the military air forces of the USSR. There can be little doubt that in a future emergency this would happen again. The aircraft of Aeroflot would be added to the total strategic air-lift resources of Soviet armed forces. It is for example of some significance that the GVF (initials stand for civil air fleet) has since the war been headed by senior air force officers including Chief Marshal of Aviation Zhigarev, with chief deputies of the air rank of Marshal of Aviation or Lieutenant-General of Aviation. About a year ago General Loginov, previously head of all Soviet Air Force tactical aviation became chief of the GVF civil air transport administration. And Aeroflot administration is staffed by a large number of reserve air force officers with experience in military flying behind them.

The beginnings of Soviet civil aviation were not so grand. The first regular air line in the USSR was a joint endeavour organised by a defeated Germany and the battered, dislocated Soviet Union. It was called Dereluft, and was founded in 1921 mainly with German capital and German personnel. It operated between Berlin and Moscow and gradually extended in the 1920s with new routes to Leningrad and a number of the main German and Polish cities. The Soviet Council for Civil Aviation was established in February 1923 and created three civil air authorities to develop passenger and other air transport. The most important was Dobrolet (Society of the

Voluntary Air Fleet) and it had absorbed the other two within a few years. They were the Ukrainian civil air authority, Ukrvozdukhput, and the Trans-Caucasian authority, Zakavia. The merger occurred in 1930 and was not surprising, for at this stage neither the number of equipped airfields nor of passenger planes could justify three separate civil air authorities. Like civil aviation in other countries, Dobrolet was still in the pioneering stage. There were great plans to expand passenger air lines, to explore the Arctic air routes, to help Soviet agriculture and to try to get transport aircraft to overcome some of the current deficiencies in the Soviet transport system, deficiencies which still exist. There are for instance estimated to be less than 200,000 miles of hard-surface motor roads in the USSR. Most of Soviet motor transport is still concentrated in European Russia; the immense area from the Urals to the Pacific has still little road transport and relies heavily on the seasonal use of rivers and of course the railways.

Up to 1930 air transport could do little to overcome the main deficiences of the Soviet transport system and this continued to be relatively true for the next twenty-five years. There was, however, individual enterprise, initiative and heroism. Seals were flown from the Caspian Sea to stock the sea of Aral, and Vopdopyanov with a single aircraft at his disposal began exploring the possibilities of establishing an air line in the north-east between Khabarovsk and Sakhalin.

At the end of the First Five Year Air Plan, which gave little or no priority to the production of transport or passenger planes, a new organisation was set up in 1932 called the "Main administration of the Civil Air Fleet". The Russian initials abbreviate to GVF and the organisation is sometimes referred to as Aeroflot. Both labels and the organisation have survived to the present day. In the twenty-five to thirty years of its existence there has naturally been a story of continuous growth. But the post-war development has been more impressive than the pre-war, and the last five years more impressive than the previous ten years. All the same pre-war expansion was not unimportant, though certainly much less than in, say, the United States, France, Germany or Britain.

The pre-war period brought no continuous policy of making enough transport aircraft for either military or civil purposes to meet the needs of either the expanding GVF Aeroflot air lines or of the military airborne forces. The first battalions of parachutists had to compete with the bomber forces for supplies first of the twin-engined TB 1 bomber designed by Tupolev and later for his four-engined TB 3. Tupolev did design the 9-10 seater three-engined passenger plane the ANT 4 which was a civil derivative of the military bomber the TB 1. In 1929 it flew from Moscow to New York. And seven years later, in 1936, another of his long-distance passenger planes, the ANT 25, helped to pioneer the Arctic air routes in a flight from Moscow to San Jacinto in California. But these were in the nature of propaganda demonstrations intended to impress people at home and abroad. It was to take another twenty to twenty-five years before Aeroflot could seriously consider competing with Western civil air lines in overseas international passenger air routes.

Apart from Tupolev, Kalinin, who disappeared in the Stalinist purges of the late 1930s, was the chief designer of Soviet pre-war transport aircraft. His eight-passenger K 5 single-engined transport plane was one of the most popular in the early years of Aeroflot when the GVF was mainly concerned with the expansion of air passenger services in Western Russia from Moscow to such places as Gorki, Kazan, Kiev, Kharkov, Rostov and Stalingrad.

Various sets of statistics have been published to illustrate the pre-war development of Soviet civil aviation and the writer is always at the mercy of conflicting Soviet sources on this or most other subjects concerning the USSR. It would seem that the overall length of Soviet internal Aeroflot air lines increased from some 20,000 to 25,000 miles in 1932 to over 50,000 miles in 1937 and to 80-90,000 miles by 1939-40. The number of passengers carried during the pre-war period went up from 25,000 to 39,000 in 1932 to 175-200,000 in 1937 and to nearly 300,000 in 1940. There was an even more striking increase in freight carried; from a total of 400-500 tons of mail in 1932, the figure went up to over 6,000 tons in 1937 and over 7,500 tons in 1940. Other freight carried increased from less than 500 tons in 1932 to nearly 36,000 tons in 1937 and over 43,000 tons in 1940. Stalin's claim at the 18th Party Congress in 1939 to a tenfold increase in civil air transport lift in the period 1933-38 was justified and confirmed by original Soviet documents subsequently captured by the Germans.

Comparisons between Soviet civil air line achievement and that of other countries is bound to be misleading. However, statistics indicated that even back in 1935 Soviet civil air lines carried much more freight than those of any other country. But this was because there was a deliberate policy of carrying freight rather than passengers. Thus in 1939 Soviet civil air lines hauled roughly ten times as much freight as U.S. air lines but the latter six or seven times as many passengers.

One of the main reasons for the delay in Soviet attempts to break into the competition for passengers on international air lines was the failure to develop any numbers of successful long-range multi-engined passenger transport planes in the pre-war period. Tupolev had designed the eight-engined ANT 20 Maxim Gorky air liner which was first test-flown in 1934 but crashed in an air collision the next year and never went into series production. His other long-range five-engined transport machine the ANT 14 was also designed in this period, but it was test-flown at a time when the Soviet aircraft production policy was geared to maximum output of single-and twin-engined fighters and bombers. With the mounting threat from Japan and Germany, it was felt that to make large numbers of tactical support planes was the first consideration.

The TB 7 four-engined machine designed in the immediate pre-war period had all the makings of a good long-range bomber or transport plane; but it only went into very small-scale production. It has been noted that the GVF civil air lines of the pre-war, and indeed the immediate post-war periods, were more closely linked to military transport operations in the USSR than the civil air authorities in other countries. Perhaps one of its most significant pre-war achievements in the light of post-war strategy, was the development of Aviaartika, an air-line which has developed polar aviation in the USSR. Since the 1920s Soviet explorers and scientists have made use of aircraft to further their work of discovery and exploration in the polar and Arctic regions. In 1929 it was decided that the Soviet civil transport aircraft flying in those areas should be put under the general authority for that area, Glavsevmorput, i.e. the main administration for the Northern Sea route. But in the crude formative years of the formation of Soviet polar aviation, few aircraft and little modern equipment could be spared for the Arctic operations of Aviaartika. There were high-lights in the pre-war period which included the air rescue of the survivors of the wrecked ship *Chelyushkin* in 1934 and the air transport supply operations for the North Pole station in the winter of 1937-38. Certainly there was also a Five Year Plan for Polar Aviation which called for 60,000 hours of flying in the 1933-37 period. The plan was fulfilled, but the target of less than three hours' per day flying on the average reflected the many problems of repair facilities, fuel shortage, and the establishing of navigational aids. There was also the general problem which afflicted all Soviet air transport operations in the five years which preceded the Germano-Soviet War, i.e. the use of more than twenty types of transport and communications aircraft, for many of which there was a shortage of spares and replacement parts.

Soviet civil aviation has also played an important part in the development of Soviet agriculture and forestry work in both the pre-war and post-war periods. From the late 1920s onwards Polikarpov's biplane utility two-seater, fitted with a five-cylinder air-cooled engine was widely used in this work, as well as for air ambulance operations. Both of these activities were the concern of the GVF civil air authority. The aircraft were used mainly to fight insect pests or crop and plant diseases. As far back as 1925 it was claimed that Soviet planes had sprayed or dusted over 6,000 acres of locust-breeding land, and the GVF figure for air agricultural operations is stated to have increased to over one million acres by 1940. The spraying of crops with chemicals, fertilizers, herbicides and insecticides became even more important in the reconstruction of the late 1940s after the Second World War, when the agricultural operations of GVF planes covered several million acres and helped to fertilize grain crops and increase the yield of rye, wheat and rice as well as to speed up the ripening of cotton and flax plants.

Even before the Great Fatherland War—as Soviet military literature calls the Germano-Soviet campaigns—the GVF was being incorporated gradually into the structure of the Soviet armed forces. When the joint ADD bomber transport command was created under General Galovanov in the spring of 1942, it included a proportion of the existing transport aircraft of the GVF civil air lines. The first major operation of Hitler's war undertaken by *Aeroflot* planes was the evacuation of aircraft and other military production plants from factories in Leningrad, Moscow, Kiev, Voronezh and other manufacturing centres of Western Russia to the safer confines of Siberia, the Urals and other parts of Asiatic Russia. In addition over 100 GVF transport machines were used in the wartime operations of the VDV command, i.e. the airborne and parachute troop headquarters.

But the main military work was done in dropping food, ammunition and supplies to Soviet and Jugoslav guerrilla and partisan forces. Estimates of the total number of transport aircraft sorties in these operations ranged from 30,000 to 40,000 for the period 1941-45. The Soviet authorities claim that the figure was 40,000. Whatever the statistical truth may be, GVF planes played a major rôle in the logistic air support given to the partisan and guerrilla movements. They also carried out special air lift operations in support of beleaguered cities invested by Wehrmacht forces such as Leningrad, Odessa and Sevastopol and they worked closely with the ADD in the special air lift which ferried troops and flew fuel, ammunition, spares and food to the Red Army in the Stalingrad winter campaign of 1942-43.

The aircraft used in both civil and military transport air operations were mainly of U.S. Douglas design, either the Lend Lease U.S.-built C 47 or the Soviet-built equivalent the Lisunov 2, also called the PS 84. It is of some interest that the word Douglas in Soviet Air Force parlance came to mean a large transport aircraft, though the C 47 and PS 84 were only medium-sized twin engined transport planes. Apart from these, wartime use in air transport rôles was made of a few obsolescent TB 3 four-engined bomber planes and the IL 4 twin-engined Ilyushin bomber, as well as the ever faithful U 2 biplane with its limited range of 300 miles and its restricted load of less than 500 lb.

In their *Great Soviet Encyclopedia* published in 1951 the USSR claimed that in the 1941-45 period the GVF carried 2,300,000 passengers and 300,000 tons of freight. While these figures are clearly a great advance on the totals for the preceding four or five years, they also reflect an inevitable slowing up in the pre-war rates of increase in passenger and freight transport. This slowing up was due primarily to the military commitments of the GVF in the Germano-Soviet War. In that war the Soviet General Staff had realised fully for the first time the importance of a large air transport command to make its ground and air forces as mobile as possible and to make the scale of parachute and airborne troops operations commensurate with the size of the cadres available to operate. The fact is that although the USSR achieved the maximum effort from the air transport resources of the GVF and ADD in the Second World War, there were a number of occasions, both in the desperate retreats of 1941-42 and in the spectacular advances of 1943-45 when stronger forces of transport aircraft would have lessened the disasters and increased the tempo of the triumphs. The USSR probably had a force of less than 2,000 transport aircraft of all types, including single-engined communications aircraft, at the end of the Second World War.

The first post-war Five Year Plan covering the period 1946-50 called for expansion and restoration in the passenger air lines of Aeroflot and in the military transport aircraft for the VDV airborne forces. But in the new jet age little attention was paid initially to modernisation with jet transport aircraft. In the 1945-50 period the passenger air line service network was extended from a length of less than 100,000 miles to some 150,000. Part of this route expansion was to Arctic areas from Archangel to Novaya Zemlya and Franz Joseph Island. In the main, however, it consisted of a strengthening of the East-West air communications by a series of new passenger services, e.g. from Moscow to Novosibirsk or the longest internal air route in the USSR from Moscow to Sakhalin island via Khabarovsk and Vladivostok, a distance of over 5,000 miles. The air lines under the Central Administration for Northern Territories (Aviaartika) were greatly strengthened and the intention was to make them the first post-war group of Soviet air routes to be fitted with long-range four-engined transport planes. A transport version of the Tu 4 four-engined bomber, the Tu 70, was put into service but as far as is known, less than a dozen were produced. But the military air transport arm operating to Arctic air and rocket bases was naturally to become a major facet of Soviet air policy in the 1950s.

Only three types of new transport planes emerged in the 1945-50 period, to be manufactured in large numbers. They were Antonov's An 2, a single-engined biplane with a maximum speed of about 155 m.p.h., a maximum range of less than 75 miles and a pay load of less than two tons. This was designed in 1947 and was intended to replace Polikarpov's Po 2 trainer. Between 2,000 and 3,000 of them have probably been built since 1950. To Western aircraft designers it may not seem to be of modern design or of very high-grade performance, but it is a striking example of Soviet production policy which tends to build aircraft of simple construction and use them for as many purposes as possible. Thus the An 2 has been used by Aeroflot on small local feeder lines on which it carried from 8 to 12 passengers or equivalent freight. It became the basic aircraft for post-war parachute training operations and in this rôle carried up to 14 troops with equipment. A floatplane version of it has carried out thousands of supply operations to Soviet meteorological and radar stations in the Arctic, and to the increasing number of bomber aircraft missile bases in this area. There is also an air survey version of the AN 2 fitted with two cameras which has done valuable geological and mapping photographic work, especially in the Urals and over Siberia. Finally there is the agricultural version, which is used for spraying insecticides or fertilisers and sometimes carried special detection equipment for geological survey work. Another aircraft similar in agricultural purpose which has been extensively used in the post-war period is the YAK 12, a four-seater aircraft which was already in quantity production in the USSR in 1945, but which was both re-engined and re-designed in the 1945-55 period. The 145 h.p. engine of the immediate post-war was replaced by a more powerful 240 h.p. engine in 1952 and the mixed construction of wood and metal was replaced by an all-metal machine in 1955. Yakovlev also designed and produced a twin-engined transport plane, the YAK 16, with a payload of about one and a half tons which represented one of the post-war Soviet failures to export passenger aircraft to Western countries. Following its first public appearance in the spring of 1948 at the International Trade Fair at Posnan attempts were also made to sell the YAK 16 to a number of Latin American countries. But the price quoted was rather high, the after-sales service conditions were not acceptable and the performance of this transport plane had little to recommend it in the late 1940s. With a maximum speed of less than 200 m.p.h. and a range of less than 700 miles it was not even an attractive proposition for the civil air lines of the Communist allies of the USSR. Indeed it faded out of service with Aeroflot air lines in the Soviet Union after two or three years and must be accounted one of Yakovlev's relative failures, just as his twin-engined transport designs of the Second World War, the YAK 6 and YAK 8, had been.

Perhaps the most successful of the post-war transport aircraft designers was Serge Ilyushin, who had worked closely with Tupolev at TSAGI since the 1920s, at first on glider design, but later on the Sturmovik tactical support planes which were so successful in the Germano-Soviet War. He had in his early years of design apprenticeship been closely associated with Tupolev's transport and bomber aircraft and at the end of the Second World War he took over

Lisunov's task of making a more modern transport plane to succeed the Soviet version of the Dakota, the LI 2. The IL 12 was the result. Although it went into large-scale production and was more extensively used in Soviet military transport units and Aeroflot air lines than any other machine, its performance was modest by the contemporary standards of the late 1940s and early 1950s. It carried 24 passengers, or up to 40 parachutists or airborne troops with equipment, or about four tons of freight. It had a maximum speed of just over 250 m.p.h. and a range of 750-800 miles. It might be compared with the American Convair 240 or the British Viking passenger machines. An improved version of the IL 12, the IL 14, appeared in 1953 and is still in production. This weighed less and had more powerful 1,900 h.p. engines. A year or so later an improved version of the IL 14 was seen with a longer fuselage accommodating up to 32 passengers. The maximum freight load was over 4½ tons and the maximum range with reduced load had been increased to over 1,300 miles. In the 1945-55 period the Ilyushin 12 and 14 twin-engined transport planes were in large-scale production not only in the USSR but also in Czechoslovakia. Ilyushin received a Stalin prize for these designs, and the civil air lines of Poland, Czechoslovakia, China and other allies of the USSR made extensive use of this moderate-performance machine.

With the decision not to build either the four-engined Ilyushin 18 or the Tupolev 70 long-range transport planes in large numbers, it looked in the middle 1950s as though the USSR was going to lag behind the USA in the vital field of strategic air lift and make no effort to compete seriously with the West in the development of high-speed long-range passenger jet transport planes. Estimates of the total transport aircraft available in the USSR during this period ranged from 3,500 to 5,000. But these totals included no jet transport planes and only a handful of four-piston-engined machines. It looked then as though the ten divisions of the VDV airborne troop command and the expanding Aeroflot passenger air lines would fall badly behind the West in their aircraft equipment. The twin-engined Ilyushin planes took nearly fifty hours to fly from Moscow to Vladivostok and nearly twenty hours from Moscow to Frunze in Siberia. By 1952 Aeroflot had expanded its air services to over 700 Soviet cities but the air fares remained high and foreign travellers and Soviet government officials occupied most of the passenger seats.

The emphasis was however still on carrying freight such as electrical equipment and machine tools to the Far Eastern industrial areas, fruit and vegetables into Moscow from the Caucasus and the Crimea and the delivery of newspapers, magazines and books printed and published in Moscow.

There was also a great programme of airfield development launched in the post-war period. But a high proportion of airports was covered with grass and used steel mats as runways, which made the continued use of light transport aircraft such as the IL 12, YAK 16 and AN 2 essential. Another handicap was the lack of blind-landing and night-landing facilities which limited many of the Aeroflot passenger and freight services to daylight operations only.

But the 1950s, and in particular the 1955-60 period, have witnessed almost a revolution in both military and civil air transport. When in 1955 General Zhakarov announced that the USSR would begin using jet transport planes, few people believed that the deputy chief of the GVF should be taken seriously. But when in 1959 Marshal of Aviation Zhigharev announced that the USSR was aiming at all jet passenger services in the 1960-65 period his announcement was taken with very few grains of salt. In the last five years, Soviet technical progress in airport equipment and in the construction of a growing fleet of jet military and civil transport planes has been truly remarkable.

The TU 104 twin-jet passenger plane was the first of the new group of long-range transport aircraft to go into regular service. After successful trials in 1955, it made a flight from Moscow to London in March 1956; and by the autumn of that year more than 20 TU 104 passenger planes were in service on Aeroflot air lines from Moscow to Tashkent, Leningrad and Irkutsk, and subsequently these planes came into service on scheduled services to Prague, Pekin, Copenhagen, Vladivostok and Khabarovsk. The TU 104 owed much of its basic structural design to the TU 16 twin-engined military bomber having the same wings, undercarriage, engines (but slightly derated) and tail unit. This is no doubt why its development and production were so rapid in the 1955-56 period. Its normal cruising speed is about 500 m.p.h., with a range of about 1,000 miles, carrying about 50 passengers. Since 1957 more modern versions have been built with modified AM 3 engines giving greater thrust and lower fuel consumption. These have been called the TU 104A and Tu 104B and can carry 75 passengers for stage distances of 1,500-1750 miles. In the autumn of 1957 the Tu 104A established new international records for freight-carrying planes in its category, including a load of about ten tons for over 600 miles at an average speed of over 620 m.p.h.

Although the Tu 104 is perhaps the best known of the new Soviet jet transport planes it is by no means the most important in terms of Soviet strategic mobility. Three new four-jet turbo-prop. transport machines—the Tu 114, the Ilyushin 18 and the Antonov 10—share this rôle. When used as civil passenger planes they are called respectively the Rossiya (Russia), the Moskva (Moscow) and the Ukraina. But their long-term importance may be as military transport aircraft. The Antonov 10 first flew early in 1957, though a twin-engined predecessor, the An 8 had been developed a year or so before this. Both the An 8 and An 10 are in large-scale production in Ukrainian factories for use in Soviet military transport units. The twin-engined An 8 carries about 40 troops with their equipment and the An 10 carries between 75 and 100 men. The An 8 has a maximum range of about 1,500-1,750 miles and the An 10 a range of over 2,000 miles. Both can operate from small grass airfields and are likely to form the backbone of Soviet military air transport unit equipment in the next year or so. The An 10 will also be used as a passenger air liner mainly on internal routes in the USSR.

Perhaps the most successful all-round machine of the new group of Soviet jet transport planes will prove to be the Il 18 Moskva. It made its trial flights in 1957 and first came into Aeroflot service in 1959, the year it was on show to the public at the International Air Exhibition near Paris. It is a good example of the great progress in passenger comfort which has been part of the new Soviet philosophy of air travel in the jet age, in which USSR air lines are competing with the air liners of other countries on international air routes. Such gadgets as individual reading lights, call buttons and ventilators, extra toilets and a bar and passenger tables have become as typical of Soviet jet air liners of the last few years as they were foreign and rare in the early 1950s. The Il 18 has a maximum range of over 3,000 miles, can accommodate 75-100 passengers and fly at an average speed of over 400 m.p.h. It can operate from both grass or concrete runways and may become the first successful Soviet air liner in the world's export market.

In size the Tu 114 Rossiya is the most impressive. Derived in part from the Tu 20 four turbo-prop bomber, the Tu 114 is at present the largest jet air liner in the world with a maximum load of over 200 passengers and the ability to fly over 100 passengers non-stop from Moscow to New York in about 12 hours. It has gone into general service on Aeroflot air lines inside the USSR and to Communist Europe. It is surprising perhaps that it is not yet scheduled for any of the main routes to the Western world which Aeroflot has been building up since 1955 when the USSR felt with some justice that her jet air liners of the next five years could compete with those of the West. Until then, Soviet expansion of air lines outside the USSR was cautious and virtually confined to Sweden and Finland. But in the last few years Soviet jet passenger planes have been proudly displaying the Aeroflot flag on routes to Paris, London, Copenhagen, Delhi and Cairo, and it would be surprising if more foreign capital cities were not added to its external services before this text is in print.

Incidentally the USSR is also developing a four-jet passenger air liner, the Tu 110. Fitted with four Lyulka turbo-jet power units each giving a thrust of about 12,000 lb. the Tu 110 can carry 100 passengers at an average speed of over 500 m.p.h. for a distance of over 2,000 miles. Although tested in the same period as the four-jet turbo-prop. Ukraina and Moskva, the Tu 110 is not yet known to be in service either on Aeroflot passenger air lines or in military air transport units. Perhaps it will not go into large-scale production. Perhaps Tupolev's design bureau will concentrate instead in the next year or so on the development of a supersonic transport version of the new Bounder bomber.

Soviet technical progress in the design and operation of large numbers of modern long-range jet transport aircraft in the past two or three years has a number of aspects of general strategic importance. There is in the first place the growing Soviet potential for capturing part of the international passenger and freight traffic from Western air lines and of selling modern jet passenger planes to countries which for the past two or three decades have been buying mainly British and U.S. air liners. The failure to sell the Yak 16 and Ilyushin twin piston-engined air liners in the Latin American and Asian markets in the early post-war period is likely to be replaced by a new period of commercial success in the 1960s with the new crop of turbo-prop and turbo-jet air liners, once a satisfactory after-sales service scheme is produced and the climate of East-West politics continues to improve as it surely will. From a military point of view two further points seem worth making. The diversion of so much bomber potential to air transport production in the late 1950s is a strong indication of Soviet satisfaction with their long-range missile production programme and at the same time a pointer to the fact that they do not propose to expand their ADD long-range bomber force greatly beyond its estimated 1955-60 strength of some 1,500 long-range bombers.

But the important point is the great improvement in Soviet military mobility in an age of atomic warfare and greater dispersal of air and rocket bases, which makes the expansion and modernisation of the air transport arm so crucial to the effective prosecution of war. In no area is this more important than in the Arctic where the Aviaartika air routes have been expanded again in the past two or three years, facilitating the exploitation of the new industrial wealth of Siberia, giving air lift support for the new series of scientific observation posts near the North Pole and providing mobile air laboratories at the Arctic camps. But of more crucial military significance is the increased air lift to the bomber bases of the ADD command, to the new interceptor fighter and defensive guided-missile bases of the PVO command and to the ICBM missile bases in the Arctic. The strategic significance of the polar air routes between the USSR and the USA is now so apparent as to be almost hackneyed. The ability to supply her air and rocket forces in this region is a major aspect of the deterrent value of the USSR's long-range attack forces, the location of which in this area must be a difficult if not an insoluble problem for Western intelligence.

But the growth of air transport facilities in the past decade has improved Soviet local tactical mobility as much as it has the overall strategic mobility of Soviet air power. Much of this has been due to the development of helicopters in the early 1950s before the emergence of the long-range strategic jet air transport resources. The most important of these in terms of lifting troops and equipment are the twin-engined Yak 24 and the twin-jet engined Mil 6. The Yak 24 can carry about a five-ton load which means up to 40 troops and their equipment or alternatively vehicles, including mobile anti-aircraft guns and jeeps. When it first was used in Soviet airborne operational manœuvres in 1954-55 it was the largest helicopter in the world in general service and production. In the next four or five years some hundreds of these YAK 24 helicopters are thought to have been added to the military air transport resources of the USSR and it is only in the last year or so that the YAK 24 has been known to be used in passenger air services of Aeroflot, e.g. in the Crimea. Smaller than the YAK 24, but in service in air transport units for a year or so longer, the MIL 4 helicopter carries only 14 equipped troops or alternatively anti-tank guns, a small jeep or two motor-cycle combinations. It has been in quantity production since 1953, but may soon be phased out of Soviet air transport units by the larger and more modern helicopters. Of these the MIL 6, which completed successful test flights in 1957, is the world's largest helicopter and likely to become one of the major means of Soviet tactical air mobility in the next year or so. It can carry 70-80 passengers and probably over 100 armed troops, for it can lift a payload of well over twelve tons. It probably has a range of well over 350 miles and its two-shaft turbine Solovev engines give it a speed of well over 250 m.p.h.

Of the helicopters used for Aeroflot agricultural and survey work the MIL 4 is perhaps most frequently encountered on crop-spraying or forestry survey work. Like other countries the Soviet Union is not yet convinced that the helicopter is the best vertical take-off or short take-off aircraft design if one wishes also for maximum performance in the air. They are therefore developing not only short take-off machines such as the ANTONOV 14 designed for air transport and passenger work of this kind but all also working on jet engines which give both forward and vertical thrust.

Despite the great modernisation and expansion of the Soviet military and civil air transport arm in the last five years, there is still much to be done before the general standards of Soviet air transport equipment equal those in the U.S.A. As late as 1957 the Soviet LI 2, the equivalent of the twin-engined Dakota, was still being used on Aeroflot air lines and even its replacements, the IL 12 and IL 14, would be obsolescent by U.S. standards. Both the IL 12 and IL 14 and the Antonov 2 were still fairly commonly met with on local Aeroflot air lines and in Soviet military air transport units up to the end of the Fifth Five Year Plan in 1955, because so many military airfields and civil airports did not have the runway and blind-landing facilities to take larger and faster aircraft. Since 1955 modern airport facilities have been developed at the more important industrial centres such as Kazan, Omsk, Sverdlovsk, Irkutsk and Khabarovsk. The Sixth Five Year Plan begun in 1956 included the construction of over forty new air terminals at such industrial centres in the USSR. These new modern airports are fitted with good runway lighting and instrument landing facilities, and as radar is introduced as standard equipment, all-weather all-the-year-round night-and-day operations will become the rule.

The Soviet Union has not yet caught up with the U.S.A. in numbers of long-range four-engined transport aircraft and it is still anxious to acquire some of the latest British-made navigational aids for operating passenger planes. But Soviet pride need no longer suffer when its latest jet transport machines are parked side by side with those of rival nations, European or American. Soviet progress in its air transport arm has in the last five years broadly speaking caught up with the West. It is therefore not surprising that since 1955 there have been bilateral agreements to operate new jet Aeroflot services to London, Copenhagen and Paris. It would not be surprising if Aeroflot operations were expanded on a world-wide scale in the next few years with the Soviet Union joining the International Air Transport Association and so announcing her full challenge to the West in the field of civil passenger air transport operations.

CHAPTER 12

SOVIET AIR STRATEGY

STRATEGY has been defined as the "art of projecting and directing the larger military movements and operations". The definition was made more than a decade ago, but it seems to have survived the fantastic revolution in military naval and air weapons which has since taken place. In the final analysis the elements which constitute the amalgam of Soviet army, navy or air force strategy must be similar if not identical to those of other greater powers. They are primarily the geography of the Soviet Union, its size, soil, climate, products and means of communication, the prospective opponents and allies of the USSR, the technological state of weapon development, the state of Soviet manpower and finally, though not very important in practice, Soviet military views on contemporary and past military doctrine, propounded by such thinkers as Douhet, the Italian air strategist, Mahan, the American doyen of naval strategy or Clausewitz, perhaps Germany's greatest military thinker on pre-atomic land warfare.

There is also the important question of the influence of Soviet political leaders and the Party on the strategy of the armed forces of the USSR. In another chapter, dealing with political influences in the Soviet Air Force, it is argued that Stalin, Khrushchev, Bulganin and others may make many pronouncements on defence matters, but ultimately exercise little continuing day-to-day influence on the Soviet "art of directing and projecting the larger military movements". In making or planning treaties with Hitler, Tito, Nasser or the Warsaw Treaty powers, clearly the Kremlin leaders must exercise an important effect on long-term naval, military and air plans. In the choice or purging of Soviet military leaders they probably come closest to affecting strategy in detail. Thus the elimination of men such as Alksnis and Tuchachevsky in the purges of the 1930s contributed to the delay in creating a force of modern long-range bomber and transport aircraft. The post-war imprisonment of Chief

Marshal of Aviation Alexander Novikov and his premature retirement after his release from prison must have affected not only the calibre of Soviet air leadership, but also the doctrines of post-war leadership to some extent. Marshal Novikov was not the only post-war victim of Stalin's irritation. His war-time deputies, Marshals of Aviation Astakhov, Kudiakov and Vorozheikin were also prematurely retired by Stalin's orders in the late 1940s, as was also Colonel-General Repin, chief of the post-war technical air services. Later came the dismissal of Marshal Zhukov by Nikita Khrushchev. But, again, it is hard to find any concrete evidence that the Soviet art of projecting and directing air and rocket operations would have been any different had these and other air leaders survived the displeasure of the politicians and kept their jobs.

Indeed, the events of the past forty years, both in war and peace, have shown that Soviet air doctrines, plans and strategy have been eminently practical and based on the stern realities of Soviet military commitments and weapon, equipment and personnel potential, rather than any personal *ex cathedra* assessments either by the Kremlin political leaders or by foreign military theoreticians. One doubts, for example, whether it could be shown that Stalin as a Defence Minister and Supreme Commander had more influence on the Soviet direction of war in 1941-45 than did Sir Winston Churchill on the British direction of war. It is true that Stalin took a leading part in the reconstitution of the Soviet long-range strategic air force, the ADD, in the spring of 1942. But in the decision to concentrate much of the British war effort on the long-range bombing of Germany in the same period, who could say if Churchill's influence was not greater?

Hanson Baldwin, the *New York Times* military editor, has written: "Strategy, like diplomacy, is the art of the possible." This dictum seems to fit Soviet air and military strategy like a glove. Soviet air plans are based on the view that first things come first. Faced by the German onslaught of June 1941 and with centuries of experience of land battles, the bulk of Soviet air power between the 1917 Revolution and the launching of Operation Barbarossa in June 1941 by the Wehrmacht naturally consisted of tactical air-ground support regiments, brigades, divisions, corps and armies. Even a large part of the naval air forces was equipped and trained to carry out this type of air operation and the long-range bomber forces also reverted mainly to this rôle in the Germano-Soviet war.

There was a time in the early 1930s, when it looked as though Soviet air strategy was to include as a major tenet the conception of long-range strategic bombing against industrial target systems. It would seem that Marshal Tuchachevsky encouraged such ideas amongst some of the senior members of his air staff such as Generals Golovanov and Khripin. Tupolev, primarily a designer of multi-engined bomber and transport planes built a force of twin-engined, so-called heavy (TB 1) bombers and later a force of several hundred four-engined TB long-range bombers. For a brief period in the middle 1930s it seemed as though the Soviet Air Force would be the first to have a heavy bomber force capable of carrying out long-range strategic operations. But as the 1930s came to an end it was apparent that the Soviet policy of putting nearly all their aeronautical eggs in the tactical air basket was predominating. No doubt the views of Douhet, Trenchard, Smuts and other strategic air power protagonists had been carefully studied and considered; but stern Soviet practicality probably intervened when it was realised what a drain on aircraft production resources the creation of a large heavy bomber force would involve. In the 1930s the USSR was sternly engaged in bringing total aircraft output up to and beyond the manufacturing levels of Germany and Japan, the major prospective enemies. In addition, Soviet aero-engine design, blind-flying and navigational equipment lagged behind the basic requirements of a modern efficient strategical bomber force. The USSR simply did not have the means of producing large numbers of heavy bombers and at the same time of fighters, light and medium bombers and Sturmovik ground attack machines to support the large number of ground divisions which the Soviet Union had to deploy as the lynch-pin of her war strategy in the 1930s and 1940s. The Soviet decision to slow down the production of four-engined aircraft was wise and practical, both in prospect and in the light of subsequent events. As a preparation for the air commitments of the Second World War, the engineering and design skills of men such as Tupolev, Petlyakov and Ilyushin were concentrated on the production of single-and twin-engined tactical air support machines such as the PE 2, IL 2, IL 10, IL 4 and TU 2. And of course the production of single-engined fighters to ensure maximum local air superiority quite rightly dominated Soviet aircraft production plans. Had Germany taken this particular leaf from the Soviet air book, the Battle of Britain, in which the Luftwaffe had an inadequate single-seater fighter force to do the job in hand, might well have gone in Germany's favour and the history of Europe would perhaps have been re-written.

Soviet achievements in designing and producing long-range aircraft in the post-war period are a clear indication that the pre-war aircraft production policy was deliberate and only partly conditioned by lack of skills. Comparison with the German aircraft industry and German air policy is probably only partially valid, but it may be useful to record that in the 1930s German aircraft designers such as Focke-Wulf, Dornier, Junkers and Heinkel produced successful prototype four-engined bombers none of which ever went into large-scale production. (The F-W 200 was never manufactured at the rate of even a dozen a month.) And who could blame Stalin and his air leaders of the pre-war period for concentrating on the creation of a large tactical air force at the expense of strategic air power? The results achieved by Soviet bombers in the wars in Spain and Finland against strategic targets such as communication systems were extremely disappointing for the Soviet High Command. It can of course be argued that this was inevitable. The Soviet bomber forces were inadequate both in the type of bombs carried (usually 250 and 500 kg.), the blind-flying and navigational skill of the crews and the strength of the forces employed. Air Chief Marshal Sir Arthur Harris called for a force of 5,000 strategic four-engined bombers to produce decisive results in the long-term air attack on Germany in the Second World War. The Soviet Air Force used less than 500 twin-engined bombers in each of the Spanish and Finnish wars, though a few squadrons of obsolescent four-engined TB 3 Tupolev bombers were also used in the Finnish campaign.

And when the full-scale air test of World War II came, in the summer of 1941 and thereafter, the Soviet need to develop strategic air forces declined as Anglo-American long-range bomber forces developed. The target systems attacked by these air squadrons, such as German aircraft industry, the Ruhr, German rail and sea communications and finally oil, were essentially the same as those which a strong Soviet strategic bomber force would have attacked. The re-creation of the ADD long-range Soviet Air Command in 1942 was a decision taken with a number of different objectives in mind. It was by no means a Soviet attempt to create a third allied strategic bomber force to join in a combined anti-German air effort. Though the Soviet Defence Ministry wanted to hit back at air targets in Germany and German-held objectives in occupied territory, there was no intention of using the new ADD long-range force primarily for strategic bombing. It was in the first place a strategic reserve air army, to be used for any air operations that the exigency of the military moment determined. It was essentially a combined bomber and transport command and used its resources at times to air-ferry men, supplies and equipment much more than to bomb at long range. As a combined bomber and transport air force it was probably without precedent in the annals of military air history.

The Stalingrad campaign showed again that Soviet air planning and strategy is at times unorthodox but nearly always essentially practical. In the autumn of 1942, when the threat to Stalingrad from the German Sixth Army became acute, General Golovanov's ADD forces concentrated several hundred transport and bomber aircraft, which in a combined air transport lift flew down to the Stalingrad area. Tens of thousands of Soviet parachutists and also crack guards rifle formations as well as thousands of tons of equipment and supplies were carried by air, and this may well have had a decisive influence on the campaign. In the succeeding months, when the tide of battle turned and von Paulus's German Sixth Army was surrounded, the German Air Force had a similar air transport rescue operation to carry out. But there was no combined bomber-transport air command in the Luftwaffe. The German air lift attempt to supply von Paulus's troops and to help them break out of the Red Army encirclement on the Stalingrad front was a failure. This was due in part to atrocious flying weather and to the strong local Soviet air superiority; but the German fault lay partly in the clumsy machinery for concentrating its air transport resources quickly at the *Schwerpunkt* of the battle. Of the dozen units of German air transport planes to be used—Junkers 86, Heinkel 111 and Heinkel 177 and Junkers 52—totalling some 350 aircraft, five were bomber units with scattered operational commitments and attached to four different tactical air corps spread out along the 2,000 mile Eastern Front. And so the German target figure of lifting 300 tons a day was never reached, although 25,000 German wounded were saved in the final evacuation. It would, of course, be wrong to attribute the collapse of von Paulus's Army to the lack of air transport, but it would appear that the Soviet Air Force had learnt a vital lesson in the Stalingrad campaign which may well have influenced their post-war air strategy' viz. the crucial importance of a strong air transport arm.

In the first year of the Germano-Soviet War, the importance of the air transport arm and of the operational mobility which it affords to both air and ground forces was a factor which the Soviet defence staff was to appreciate grimly as their forces retreated hundreds of miles eastward to Leningrad, the Moscow province, the Ukraine and the Crimea. Hundreds of unserviceable aircraft which might have been flown back to safety, much valuable air and army equipment, countless wounded and large supplies of bombs and fuel were abandoned in the face of the Wehrmacht's blitzkrieg advance of 1941, because there was no large force of air transport planes available to evacuate men and supplies. On the positive side, however, the Soviet authorities used both rail and air transport intensively to evacuate aircraft factory tools and jigs from the threatened plants in and near Leningrad, Moscow, Kiev, Taganrog and Voronezh to aircraft production centres beyond the Urals. This successful evacuation not only helped Soviet aircraft production to recover and expand at a rate which surprised both Russia's friends and foes, it also probably stimulated the Soviet Defence Ministry to recreate the ADD Bomber Transport Command in the spring of 1942.

However, despite the aircraft industry evacuation success, the Stalingrad air lift, the operation of over 250 transport planes in large-scale parachute and airborne operations and finally the extensive use of air transport to supply both Tito and Soviet guerrilla and partisan forces in Yugoslavia and German-occupied territory inside and outside the USSR, there can be little doubt that one of the lessons which the Soviet High Command drew from the Second World War was the inadequacy of their air transport resources and the vital need to expand them after the War both in the military and civil fields. Details of what they have achieved to this end are given in the chapter on transport aircraft and helicopters. In his study of the Soviet Airborne Forces (see *The Soviet Army*, Weidenfeld and Nicolson, pages 376 *et seq.*) General Student, Commander-in-Chief of the Wehrmacht's parachute divisions in the Second World War, makes the point that although the USSR had twelve first-line airborne brigades in 1941 and that these forces were greatly expanded in the war, they only operated in brigade strength on three or four occasions. He attributes this not to the quality of Soviet airborne

LAVOCHKIN, YAKOVLEV AND MIKOYAN—*c.* 1948

ANTONOV—1958

TUPOLEV—1957

RUSSIAN AIRCRAFT DESIGNERS

LT.-COL. E. BERSHANSKAYA, COMMANDER OF A GUARDS REGIMENT

GUARDS CAPTAIN MARINA CHECHNOVA

GUARDS CAPTAIN MARIA SMIRNOVA, HERO OF THE SOVIET UNION

GLIDER PILOT OLGA KLEPIHOVA

WOMEN IN THE SOVIET AIR FORCE

troops or to their equipment, both of which were high, but to radio and blind-flying deficiencies in the Soviet air transport squadrons which limited their use in bad weather, at night and even in large-scale formations in daylight. But, like other military writers who discuss the enigma of the employment of Soviet airborne forces in the Second World War, Student does not mention the numerical weakness of the Soviet air transport units in relation to their overall air transport commitments. It is true that in the Soviet advances of 1943-45 a number of airborne troop objectives presented themselves, e.g. on the Oder and in the conquest of Berlin and Budapest, the capture of which by parachutist and airborne infantry forces might have shortened the war by weeks or even months. But the same basic reason which was a major cause of the failure of the 1st Allied Airborne Army at Arnhem, the lack of air transport planes to lift simultaneously the airborne troops available, partly decided the Soviet High Command against risking their airborne forces in large-scale operations, which though tactically important were never strategically vital. And who could say that once again Soviet air strategy was not eminently practical? Was it not then politically and strategically more desirable that Tito's partisans should be able to survive against German and other forces in Jugoslavia. And if the Soviet ground and tactical air forces were sometimes held up by the lack of supplies which might have been air-ferried in the final year of the war, was it not more important to build up mass tactical air forces which would ensure local air superiority over the Luftwaffe?

Certainly, there is a case for criticising Soviet air policy in the 1943-45 period on the grounds that it exaggerated the importance of local air superiority and so built too many single-engined fighters partly at the expense of transport aircraft. This policy has gradually been modified in the post-war period. By 1950 Marshal of Aviation Skripko had been appointed head of the air transport regiments of the Soviet airborne troops and had a force of over 500 transport planes, mainly twin-engined. This force was soon to be supplemented by large numbers of MIL and YAK helicopters for lifting parachutists, troops, artillery, jeeps and supplies. In 1952 General Gorbatov became Commander-in-Chief of Soviet Airborne Forces which were declared to be a "Central Administration" independent of the air force. But quite rightly the development of Soviet transport aircraft in the post-war period took second place to the development of the strategic jet fighters and bombers and of guided missiles for both defence and attack. General Student (*ibid.*, page 382) estimated the strength of the Soviet air transport fleet in the spring of 1955 at 7,500 transport planes. Unless the figure includes civil planes used for Aeroflot passenger and agricultural purposes and parachute trainer planes such as the Antonov 2 biplane, it is likely to be on the high side; but the important point is that since 1955 the Soviet air transport arm has at last enjoyed a top priority in Soviet aircraft production plans. In the 1960s Soviet air mobility will be at least twice as great as ever before. This is because the USSR is creating a large force of modern twin- and four-engined long-range transport planes such as the Antonov 8 and 10, the Ilyushin 18, the Tupolov 114 and 104, and at the same time they have developed large helicopters to a greater extent than the West. At the moment, NATO countries have nothing to compare with the Mil 6, capable of carrying a load of over twelve tons, or the Yak 24 which has been in service for several years and can carry 40 troops or an equivalent load of guns, supplies or equipment.

It is clear that the Soviet High Command have realised that in an age of transportable missiles, potential atomic warfare, and strategic air commitments, the air transport arm must be strengthened to its maximum, so that effective dispersal and mobility of its ground and air forces in battle can be achieved. Experts write extensively about the importance of "fire brigade" forces; the Soviet too may have need of them and in just as many places. The Middle East also represents an area of disturbance and uncertainty for the Kremlin. The political promise of 1955-56 when Egypt and Syria were being equipped with Soviet planes and army and navy equipment, and Communist penetration of the Arab world seemed menacing, has been temporarily replaced by a series of disappointing checks to Soviet hopes and ambitions. Will Hungary rise in revolt again, and what about Poland? Will Outer Mongolia become a sore thumb once again? What encouragements or checks to China's advances in Asia does the Kremlin anticipate? How far must Aeroflot, the Soviet civil air line, compete with Western air lines both in securing a share of world air traffic and in selling the new Soviet jet transport planes in overseas markets? Should the USSR endeavour to do this before she has equipped the air lines of her Communist allies in Europe and Asia with the latest jet aircraft available? The appointment of senior military air leaders, such as Zhigarev and Loginov, to be in charge of Aeroflot in the last five years and of an officer of the rank of Marshal of Aviation to be in command of the military air transport arm are in themselves signs that air transport now plays a greater rôle in the priorities of Soviet strategic air planning than previously.

That this is now so, is no doubt partly because the more vital air questions of Soviet strategic attack and defence and of local tactical air superiority had already been given full priority in the 1945-55 period. In the immediate post-war re-appraisal of Soviet air doctrine and strategy it was the strengthening of the strategic air defence and air attack resources which were the subject of the first reorientations. There were two main reasons for the Soviet re-evaluation of the rôle and potentialities of air power. Stalin and his defence staff were impressed by the sight of the Anglo-American bomb damage they found in Berlin, Leipzig and other German cities, and by the evidence of the damage done by American bomber forces to Japanese cities and industry. The atomic bombs dropped at Hiroshima and Nagasaki were the most devastating and dramatic, but were in fact only a minor element in the overall U.S. strategic air attack on Japan. The spoils of Peenemunde with its long-range rocket potential also gave Stalin and his defence colleagues a new realisation of the great potential of long-range air attack. One of the more reliable Senior Soviet Air Force officers of my acquaintance has assured me that when Stalin realised in 1945-46 that there was a future for intercontinental ballistic missiles he urged his technical staffs to go ahead as quickly as possible so that he could bludgeon President Truman at the diplomatic and political levels. While at the conference table, Stalin and his colleagues showed no great surprise, enthusiasm or interest in the news from Hiroshima and Nagasaki. Both in the field of atomic weapons and long-range military rockets, the Soviet authorities gave maximum priority and technical resources to these weapons in the five years that followed World War II, a period in which the armaments policies of the Western powers included large-scale demobilisation, putting aircraft-carriers and other weapons into moth-balls and developing long-range rockets, rocket-propelled aircraft and helicopters at a leisurely pace compared with the USSR.

While it was in the 1950s that the USSR achieved technical parity or better with the West in nearly all modern air weapon and equipment fields, the foundations for this strategic offensive policy were laid in the five to ten years that followed the Second World War.

It is hard to share the views of Dr. Raymond Garthof on Soviet air strategy during this period. He finds that "the period from the end of the war until 1953 was dominated by the dogmatisation of World War II doctrine limiting air power to the rôle of supporting the ground forces". This he declares was due to the canonisation in the USSR of the Stalinist military doctrine. While it is true that the majority, perhaps two-thirds, of Soviet air regiments were trained and equipped for tactical air support during the period 1945-53, it is also true that the strategic bomber arm was reconstituted in 1946 and by 1953 had, according to General Twining, then Chief of Staff of U.S. Air Forces, reached an operational strength of 1,000 four-engined Tupolev bombers. Clearly whatever Stalin and other military leaders might say and whatever might be written in Soviet military literature, a new impetus to strategic bombing and other means of long-range air attack was given by the German, British and U.S. air developments of the 1940s. Soviet air leaders, such as Vershinin, Sudetz and Novikov may have inevitably encountered some opposition from the Soviet Army in the building up of a large long-range air command which was not primarily to support ground fighting needs. It is easy to say, as a number of military writers have, that the Soviet Union thought in terms of local wars and battering-ram tactics and glorified the artillery and infantry and the tactical air regiments that supported them; but the building of a force of 1,000 four-engined bombers in the space of 7 or 8 years, and of a training scheme in blind-flying and long-range navigational skills to support them is no mean achievement. And this was during the same period that the USSR strengthened and modernised its strategic air defences and created a much larger PVO metropolitan air defence command of over 3,000 jet fighters, with a growing radar early warning network which gradually provided air defence in depth.

It is true, as Dr. Whiting of the U.S. Air University has pointed out, that from 1945 to 1949 the Soviet Union had no effective reply in offence or defence to the strategic nuclear air striking force of the United States and that during this period (and indeed in the succeeding years) the utterance of Soviet military spokesmen and politicians tended to belittle the military power of nuclear air weapons and to idolise Stalin publicly as the architect of Soviet military science. But at the same time there was clearly a major revision of Soviet air doctrine in 1945-46 which occasionally emerged in public print. There was for example the 1946 article by General of Aviation Taterchenko in a Soviet Air Force journal which concluded that a greater rôle would be given to strategic bombing. It stated that "along with tactical aviation there must also exist strategic aviation . . . contemporary air forces can carry out not only tactical but also strategic tasks which no arm other than aviation can fulfil . . . in future engagements the place of application of the main force will be not so much the front as the rear of the enemy . . .". It is indeed the Taterchenko viewpoint that became more and more the reality of Soviet air doctrine in the post-war period, and not the trite utterances of Bulganin, Voroshilov, Taranchuk and others who created a public façade of support for Stalinist air and other military doctrines. And if in January 1955 General Grichin of the Soviet General Staff proclaimed that Western reliance on air atomic blitzkrieg was reminiscent of Goering's claims for the Luftwaffe before World War II, none the less by that period the Soviet Union were well on the way to achieving atomic blitzkrieg parity with the West in both tactical and long-range bomber and rocket weapons.

Conflicting views of Soviet air strategy and doctrine are, of course, published in the USSR. When for instance Colonel Volkov writing in *Military Thought* stated flatly that "no independent section of aviation can play such a rôle as operations conducted in the interest of ground troops", he was expressing only a partial and probably parochial view. In 1946 too Marshal of Aviation Skripko, then Deputy Commander of the reconstituted long-range bomber force wrote: "One of the means of active operations against the military potential of the enemy is strategic aviation. The threat of its massed operations alone compels the enemy to hold large forces in her air defence system, diverting them from the front." This was the view of a strategic bomber commander and was again no doubt parochial.

What has in fact happened is that both tactical and strategic air power has been developed side by side in the USSR in the post-war period. Naturally enough, the Soviet jet-engined fighters and ground-attack and medium bombers preceded the strategic bombers, which did not initially replace the conventional Tu 4 four-engined bombers until the middle and late 1950s. In this, Soviet long-range bomber policy differed from that of the United States, which developed six-engined B 47 bombers before the USSR developed its equivalent Tupolev jet bombers; but in the field of long-range rocket weapons the USSR began to develop and manufacture before the U.S.A. and that is the main reason why to-day in the field of intercontinental ballistic missile attack and the development of earth satellites and space travel, the USSR is probably a year or so ahead of the U.S.A. Long-range rockets, like helicopters, have a chequered history. Helicopters could easily have played a major rôle in both ground and air operations in the Second World War if they had not been universally underestimated as a practical air weapon for reconnaissance patrol and transport operations. The same underestimate is true of the potential of long-range military rockets. The rocket is not a European invention. It came from the Arabs, who got it from the Chinese, who used them in wars against the Mongols in the thirteenth century and perhaps before. In the eighteenth and nineteenth centuries, the British were to suffer from the attentions of Indian-launched rockets in their wars against Tipoo Sahib and were themselves to use rockets in raids on Boulogne, Copenhagen and Leipzig. But it was the German V 2 weapon which initially gave the long-range rocket its present predominant place in air strategy and the Soviet defence leaders were quicker than the American to seize both the long-term and long-range importance and potential of the new German rockets.

But the time has not yet come when rockets, both tactical and strategic, are entirely replacing manned fighters and bombers in Soviet air thinking or practice, despite the 1957 pronouncements of Khrushchev and Marshal Vershinin. In that year, the Soviet surprised the world by being the first nation successfully to test an intercontinental ballistic missile and launch a sputnik earth satellite. It was natural that Soviet military and political leaders should follow these great technological successes with public pronouncements consigning manned fighters and bombers to air museums. But Soviet air policy in the last two or three years has revealed a dichotomy which makes their present and future air strategy and planning as uncertain as that of the Western powers. Side by side with the large-scale production and development of rockets has gone the development of faster and more powerful manned aircraft in the fighter, tactical bomber and strategic bomber fields. To succeed the Mig 19 and Mig 21 and Sukhoi supersonic fighters in the PVO strategic air defence regiments, there have been in 1958-60 not only better and longer-ranged ground-to-air guided missiles but also the MIG 31 and later SUKHOI supersonic fighters, fitted with more powerful jet and rocket engines and with speeds in the Mach 3 and Mach 4 range. In the field of tactical support bombers, the IL 28 and YAK 25 subsonic planes are being followed by new supersonic Yakolev and Ilyushin ground-support tactical bombers. Finally, in the category of long-range strategic bombers, the subsonic Bear, Badger and Bison jet bombers are being followed by a supersonic bomber and the further attempt to develop atomic-engined bomber-reconnaissance aircraft.

Once again this is a realistic Soviet air policy. The time has not yet come when the liquid-fuelled intercontinental ballistic missile can accurately attack rocket and nuclear bomber bases or rocket-launching submarines. No figures about Soviet rocket-launching failures have of course been published since 1957; but the fact that a new series of solid-fuel rockets is being developed in the USSR to replace the liquid-fuel missiles, added to reports from U.S. sources, leads to the inescapable conclusion that the USSR cannot yet consider her long-range liquid-fuel rocket weapons completely reliable. The United States Secretary of Defence stated in January and February 1959 that by 1961, the USSR would probably have a lead of three to one over the U.S.A. in the field of intercontinental ballistic missiles and might deploy some 600 of them in that year. But the nub of the strategic problem is the answer to the question how many of the Soviet rockets can be accurately launched and guided to U.S. deterrent bomber and rocket bases in North America. I do not think that the Kremlin can know the answer, however confident their published statements on the accuracy of Soviet ICBM missiles may sound. It is one thing to fire sputnik earth satellites and luniks with the high-grade checking and fuelling arrangements which inevitably accompany these widely publicised events. The same standard of firing and servicing cannot be expected in, for instance, the Arctic north, partly because of storage and other climate problems, but much more important because the technical, personnel and operating standards which can be reached in the general use of a weapon in Soviet rocket batteries must be so much lower than in the show-piece demonstrations of Soviet progress in space travel. We are told by reputable scientists that the long-range liquid-fuel rocket is not yet a fully reliable weapon and this may well apply to Soviet as to American rockets.

Even more fundamental are Soviet uncertainties about strategic air defence. Some of them are indicated in the chapter in this book dealing with this subject. The reader is also referred to an interesting chapter on "The Impossibility of Effective Defence against Air Attack" in *The Failure of Atomic Strategy* by F. O. Miksche (Faber, 1959). If it is true that the USSR cannot guarantee either the defence of any part of her homeland or to subdue with certainty any given percentage of the NATO air rocket deterrent by surprise or preemptive blows, then we have, as I believe, reached a vital period of stability in East–West military strategy, which provides a few years to improve interdependence and military potential in the NATO alliance or to evolve some genuine form of international disarmament.

The West tends to talk easily of Communist aggression in Finland, Korea, Hungary and Tibet. But the aggressive military plans of the USSR have been the very soul of caution. Stalin pulled his air force out of Spain before there was trouble there on a European scale. The Soviet-Finnish War was carefully limited and localised. No one can accuse the USSR of not taking precautions to avoid major conflict with Germany in the Second World War, though clearly the Kremlin indulged in major political mischief in Poland and the Baltic States and on the Roumanian borders. And since World War II is there anything in the record of Korea, Suez, the Formosa situation or the recent events in Jordan, Lebanon and Syria to suggest that the outbreak of a Third World War is any serious part of Soviet strategic intentions?

In my view there is no military threat from the USSR to the Western world as long as the present balance of nuclear and thermonuclear terror can be maintained. Any potential threat will probably be less in the middle 1960s when long-range rocket bases become less vulnerable, particularly the submarine-launched bases which will be so scattered and difficult to detect. There is, of course, always the possibility that, as Dinerstein has said (*War and the Soviet Union*, Atlantic Books, page 204): "A Soviet political leader or faction, convinced of Soviet ability to win a nuclear war, will overestimate the indications of a possible American attack." But this element of instability in the present is a very small one indeed. It recedes with each year that passes because both Soviet and U.S. retaliatory weapons become less vulnerable to the surprise blow, with hard rocket sites, rocket-launching submarines and longer-range and more efficient early warning radar.

It is for these reasons that one concludes that Soviet strategy is "all mischief short of war" and is centred round a number of economic and political purposes aiming at the disruption of the Western world. The Atlantic powers are to be isolated as far as possible from the evolving nations of Asia, Africa and Latin America. Propaganda by radio, press and television against "Western Imperialism", capital loans on favourable terms, penetration of the export markets of the Atlantic powers—these are the main weapons of Soviet policy. The Soviet Army, Navy and Air Force are an insurance to guarantee the implementation of that policy. But even under conditions of temptation and provocation, the USSR is unlikely to throw down the glove, for it could inevitably lead to the uncertain arbitrement of nuclear war. The USSR has much to gain from peace and she is probably better prepared than the NATO powers to fight most if not all types of land or air warfare. Her tactical air forces with their greatly increased mobility could certainly achieve air superiority in most sectors of any limited local war in which the USSR was engaged. But the problem is always to know for how long the war would be limited. And in the thermo-nuclear war which followed, how could the Kremlin know soon enough what the effect of her nuclear attack was, be it surprise pre-emptive or retaliatory? Can she rely on getting intelligence from air reconnaissance, agents or earth satellites in the space of a few hours or so? The fact is that in the present state of thermo-nuclear weapons, there are no canons or air and rocket strategy to guide either side, no shore-lines of military experience to give shape to air planning. It is this uncertainty, ignorance and absence of military precedent which is perhaps the most stabilising element for peace in the thermo-nuclear age.

CHAPTER 13

AIR ALLIES OF THE USSR

THE air policy of the USSR underwent many changes in the period following the end of the Second World War. There was a greater emphasis on long-range bomber forces, the development of long-range rockets, the gradual emergence of a strategic air defence system and the expansion of the air transport arm. All these were the result of the Soviet need to have air forces commensurate with their new global commitments and to equip them with the most modern air weapons. In addition the Kremlin undertook major new political and military commitments, including that of becoming the chief source of air supply at first to a number of Communist countries and then later to some relatively uncommitted countries in Asia and the Arab World.

During the Second World War, the Soviet Air Force had been required to provide only a small proportion of its aircraft resources to equip and train the French, Polish and Czech air squadrons who fought with Soviet air regiments against the Wehrmacht from 1943-45 on the Eastern Front. But when in 1948 the Iron Curtain came down in Europe the Soviet Union became the air arsenal of the new Communist countries. She gradually had to undertake the considerable commitment of providing operational, training and civil and military transport aircraft, as well as radio and radar equipment and flying instructors, bombs, ammunition and fuel for the air forces of Albania, Czechoslovakia, Poland, Hungary, Jugoslavia, Roumania, Bulgaria, Outer Mongolia, North Korea, Northern Vietnam and especially Communist China. These were the extra supply commitments of the 1945-50 period. In the 1950s further air forces have been supplied with Soviet air equipment, notably the new air forces of Eastern Germany, created in 1955-58, the air forces of the United Arab republic, refitted with Soviet and Czech built machines and equipment during the same period, and finally the air forces of Afghanistan and Indonesia and no doubt soon the Cuban Air Force.

It is a measure of the strength of the Soviet aircraft industry that it was able to undertake these new commitments and indeed anxious to expand them further. The attempts to sell fighter planes to India and Switzerland are but two examples. While Britain and the U.S.A. ran down their aircraft industry from the peak production levels of the Second World War during the early post-war period, there is little to show that the Soviet Union did the same.

In the 1945-50 period the overall aircraft requirements of Communist countries outside the USSR must have involved an annual Soviet commitment of several thousand aircraft—perhaps between 5,000 and 7,500. Most of these aircraft up to 1950 were in the non-jet category—Tu 2 and Il 10 close-support bombers, Lavochkin and Yak 350-400 m.p.h. fighters and Lisunov and Ilyushin twin-engined transport planes. Naturally the USSR tended to provide obsolescent air equipment to her satellite allies in this period.

The major recipient of aircraft and aircraft equipment was the Chinese Communist Air Force, to-day the most powerful Communist air force apart from the Soviet Air Force. Communist China fought her civil war against Chiang Kai-shek's army and air force almost without the help of air power but nevertheless ousted them from the Chinese mainland in 1949. At the time, the Kremlin was probably aware of impending events in Korea, for the Soviet Union immediately began to pour in flying instructors, engineers, ground staff, radar equipment, military and civil aircraft, ammunition, fuel, bombs—in fact everything to speed up the process of creating a strong Chinese Communist Air Force. Never before in aviation history had an air force been created in such a short period. By 1951 General Vanderburg, then the United States Air Force Chief of Staff, publicly expressed the view that "Communist China had blossomed overnight into one of the most powerful air forces of the world". It would seem that the Soviet Air Force quickly replaced the heavy losses sustained by the Chinese Air Force during the Korean War, for in 1955 General Nathan Twining made a public statement in Washington carried by newspapers and radio stations in more than fifty countries. He said that he had informed a Congressional Committee that the air force of Mao Tse-tung was the fourth most powerful in the world. Statistically that statement was true and it is likely to be numerically true now and in the future. Indeed one can expect that some time in the 1960s the Chinese Communist Air

Force will achieve third place behind the Soviet and U.S. Air Forces. But it will probably be a decade at least before Communist China has an air force commensurate with her military political and strategic hopes as a great Asian power. This in effect means having a modern strategic bomber force with large numbers of atomic weapons or a force of long-range atomic rockets. At the moment the Chinese Communist Air Force probably has an operational strength of about 3,000 aircraft, but it lacks the long-range nuclear strike power on which strategic air strength is currently based and is essentially an air force equipped with fighters and tactical bombers.

If no other modern air force has developed so rapidly, equally no other air force has inherited initially and on the spot such a valuable aeronautical legacy, or at the same time enjoyed from a major world ally such priority in air supplies and for so long. It is now more than a decade since the Soviet Union first began to equip the Chinese Communist Air Force. For the first five years of this decade China had more modern Soviet air equipment sent her than that received by any of Russia's air allies in Europe. In the last few years, however, the rate of modernisation has to some extent slowed down. No jet bombers have so far been sent to replace the obsolescent four-engined Tu 4s, which went from the USSR to Communist China at the end of the Korean War.

In the middle and late 1950s, one wondered if the Kremlin was going on to give Communist China atomic bombs and modern long-range jet bombers, such as the twin-engined Soviet Tupolev 16 or the four-engined Tupolev 20. But it is now clear that, for the last five years or so, the USSR has deliberately slowed up her technical air help to Peking in the crucial field of long-range strategic atomic air power. It is more than five years since Molotov, then a Kremlin luminary and sitting firmly in a senior political saddle, announced for the benefit of our anxious Western world that the USSR was about to endow Peking with membership of the world atomic club by despatching Soviet atomic bombs to Communist China. But it seems that both Molotov and the Chinese Communist Air Force have since fallen from the previous high estate which they enjoyed in Kremlin councils.

What then is the present position? During the last five years, the Soviet Defence authorities could very easily have given Peking a force of modern long-range jet bombers and a supply of atom bombs and so bolstered Communist China's entire Asian strategy against Formosa and against India and Pakistan. During this period, Soviet long-range air attack has gradually changed over from relying exclusively on manned bombers, to relying more and more on long-range missiles, such as the T 2 and T 3. But for reasons best known in the Kremlin, the modernisation and strengthening of the Chinese Communist long-range bomber force have not taken place, China has received no atomic bombs or missiles from the Soviet arsenal, in which there must now be quite a few bombs to spare.

This is not to say that Soviet air aid to China has stopped. Far from it. Since 1955 more and more of Peking's aspirant pilots have been trained at Tashkent, Kiev, Vladivostok and Novosibirsk. The Chinese Communist Air Force has had hundreds of new MIG fighters, such as the subsonic MIG 17 machines and supersonic MIG 19 planes from Soviet aircraft factories, they have also had training help and transport aircraft for their parachutist formations and also a supply of Soviet MIL helicopters. Soviet equipment has expanded and modernised Chinese air bases.

But if China is to become a major world power and back her political pretensions in Asia by military might, then full and strong membership of the growing international atomic club is a necessity. What then is the present position? The Russians have built a heavy water nuclear research reactor in Peking. It is a modest affair and certainly not designed to produce plutonium for making atomic weapons. Soviet atomic engineers did not begin constructing the reactor until the summer of 1958 and they gave it an output capacity of only about 10,000 kilowatts. Thus it compares with some of the early experimental nuclear reactors built at Harwell in England. Of course the Peking reactor can make fissionable material for atomic bombs. There can be little doubt that within the next year or so China will be capable of producing an atomic weapon.

But the really important point is not that China can emulate de Gaulle's performance of exploding an experimental atomic weapon over desert territory. The crux of the matter is that the USSR has deliberately not provided China with an easy means of getting atomic weapons, either by giving her suitable nuclear reactors, by simply posting atomic weapons to Peking or by giving the Chinese Communist Air Force modern means of long-range delivery by aircraft or missiles. She has deliberately chosen for her senior Communist ally the slow and hard path of entry into the international nuclear club.

It becomes a great game of Asiatic atomic guesswork when one attempts to predict the number of atomic bombs the Chinese Communist Air Force may be able to drop within the next five years or so. Some sources think that the first Chinese atomic weapon will be exploded in 1961 or 1962. Others put the date as 1963-64. In the absence of authentic data on the training of Chinese nuclear physicists and engineers, and on the extent to which the USSR has helped China to develop fissionable material one can only resort to a rough estimate. One U.S. source says there is a second nuclear reactor in Manchuria, in addition to the Peking heavy water research reactor.

But it seems a fair conclusion that it will not be before the late 1960s that China begins to stockpile atomic bombs commensurate in numbers with her political ambition in Asia. And by then, in my view, world disarmament will be in force and China will be taking part provided two conditions are met. First the restoration of Formosa to China and the ending of the present régime in Taipei. Secondly the admission of Peking to the United Nations. Incidentally Chiang Kai-shek's forces seem to think that Communist China will have atomic weapons much sooner than I do. In June 1960 a Taipei communiqué announced that defensive exercises against atomic, chemical and biological weapons had just been carried out.

Apart from help from the USSR, Communist China inherited air bases, aircraft factories and air supplies generally from both Japan and the United States. The U.S. Army Air Force built a large number of well-equipped air bases in Western and Central China. China was intended to include a number of major bases from which the final B 29 air offensive against Japan was to be launched in 1946. Quite apart from the American Superfortresses of the Twentieth Bomber Command, there were the airfields used for the 1942-44 air lift to China and the bases required for General Chennault's China Air Task Force. And so at places such as Hengyang, Lungling, Tanchuk, Paiching and Kunming, Communist China took over the results of U.S. airfield construction in the Second World War. Japan of course had made Manchuria her chief overseas territory for aircraft production, aircraft training and air supply for her war in China. These facilities, with many Japanese pilots, engineers and technicians, fell into the Chinese Communist lap in 1949.

But despite all the legacies and the help from the USSR, Chinese Air Force operations in the 1950s have shown on the whole a basic inexperience and ineffectiveness. In the Korean War they had the backing of over 1,000 MIG 15 jet fighters, as well as Soviet-operated radar and radar-controlled anti-aircraft fire. At the time the MIG 15, apart from the lack of a radar gunsight, was roughly the technical equal of the best jet fighter planes in the world and was in fact clearly superior to many of the machines in United Nations air squadrons. More than that, the Chinese Communist MIG 15 regiments had relatively speaking a simple tactical task. They had no escorting of bomber formations to carry out. By and large they had to defend the same airfields, power stations, bridges and other targets near the Yalu river. But despite the tactical advantage of defending the same limited area day after day and the enjoyment of strong local numerical fighter superiority, they failed to do their job of local air defence. The main reasons were the inferior training and quality of the Chinese pilots and the fact that without a radar gunsight or air-to-air guided missiles the MIG 15 was a poor gun platform from which to shoot down bombers. However, since the Korean War both these weaknesses have been remedied to some extent. The supersonic MIG 19 which is now gradually equipping the Chinese Communist air regiments is fitted with a radar gunsight, and the standards of pilot and crew training have greatly improved. The Korean War was after all the baptism of fire for the Chinese Communist Air Force and their combat performances should not be judged too harshly.

Chinese Communists are themselves only too aware of the limitations of their air force. At the annual conferences of air staff officers in Peking in the years that followed the Korean War, there was general agreement that training standards must be raised and the techniques of modern air power mastered. The two targets set almost annually are the liberation of Formosa and the strategic air defence of China. Like many staff discussions they have had an ethereal air, since the plans are based on totally inadequate resources. The post-Korean-War operations and development of the Chinese Communist Air Force has been spectacular or unspectacular according to one's viewpoint. Airfield development has certainly been impressive, especially in Eastern China and opposite Formosa. New or larger air bases have been constructed near Changtung, Lungtien, Kienow, Lukian, Foochow, Swatow, Wenchou as well as near Shanghai,

Canton and Shangsha. At first airfield development was primitive. As one U.S. journalist wrote picturesquely. "The thirteenth and twentieth centuries met in the Chinese countryside." But no longer is most of the constructional work done by manual means. The modern Chinese Communist Air Force is equipped with tractors, bulldozers, and other earth-shifting machines. One must expect that in the 1960s there will be at least 250 modern airfields in use and perhaps several hundred. Thanks to technical help from Poland, Hungary as well as the USSR, blind-landing and early warning equipment has been gradually installed at many of the airfields.

Despite her many air training commitments to nearly a dozen other air forces, the USSR has provided *ab initio* and operational training facilities on an expanding scale for Chinese air personnel at Tashkent, Kiev, Vladivostok and Novosibirsk. At these air-training schools they are given the full facilities of the greatly improved Soviet air training standards of the last decade. And at other places such as the signals academy in Moscow and the engineering institutes at Voronezh, Kiev, Kazan and Komsomolsk, Chinese Air Force officers are given courses in the use of radar and radio equipment and in the design and maintenance of jet engines. So far there is no evidence available of any large-scale Soviet training of Chinese Communist air personnel in the use of guided missiles for defence or long-range rockets for attack, though since 1954 it is believed that some Soviet instruction in long-range bombing and navigation has taken place, reportedly at Kazan. This followed the Soviet delivery to China of between 50 and 100 absolescent non-jet four-engined, Tu 4 bombers in that year to the Chinese Air Force.

What are the Chinese Communist Air Force doing themselves to modernise their air force and to raise training standards? Some ten years after the creation of the first squadrons in 1949, they have designed and produced their own elementary trainer type aircraft which will be also used for communications; they are also manufacturing under licence a fair range of Soviet and other machines. Manufacture of the Soviet M 11 160 h.p. five-cylinder engine has been going on in Manchuria for at least three or four years for installation in a Peking-built version of the Antonov Ptchelka, called the "Capital". The same type of engine has been installed in the Chinese Chinko, a version of the Soviet trainer the Yak 12. A factory at Shanghai has been turning out a single-engined floatplane called

COL. A. POKRYSHKIN, THREE-STAR HERO OF THE SOVIET UNION

MAJOR G. ONUFRIENKO, FIGHTER-PILOT. HERO OF THE SOVIET UNION

JUNIOR SERGEANT M. A. KOLMEETZ

LIEUTENANT KOLTICHER, MEMBER OF STALIN "SOKOL"

SOVIET AIR FORCE TYPES

PARACHUTISTS ABOUT TO EMBARK IN A TB-3, NORTH CAUCASUS (1942)

OFF DUTY IN THE SOVIET FIGHTING SERVICES

the "Flying Dragon" (Hiryu) to be operated from China's network of canals in support of the agricultural and industrial drives in China's rural areas. At Harbin a small land-based monoplane for similar uses has been made called the "Heilun Kiang" with a modest speed of about 100 m.p.h. and a range of some 400 miles.

But China's aircraft industry can assemble and no doubt will soon be designing and manufacturing large modern jet planes. For at least two or three years an aircraft factory near Mukden has been assembling Soviet MIG 17 fighters at a delivery rate of over twenty a month and a current change-over to the supersonic MIG 19 is likely. The Soviet Antonov 2 biplane utility and transport aircraft is being built in Manchuria and also the much larger twin-engined IL 14 non-jet passenger air liner. An interesting development of the last year or so has been Chinese Communist interest in manufacturing planes designed in non-Communist countries. Chinese technical air missions have for instance been to Switzerland and Sweden with a view to getting licensing manufacturing rights for the native-designed jet planes of those two countries. At the present the aircraft industry probably has a capacity for only 2-3,000 planes a year; but expansion is in progress and it is possible that within a decade or so the Chinese Communist Air Force will be making most of its own military and civil aircraft.

It is pointed out by the air statisticians that Communist China has or will soon have after the U.S.A. and the USSR the world's largest air force; but in terms of present and future commitments it is certainly not yet the third most powerful in the world. With a first-line combat strength of 2,500-3,000 aircraft in the late 1950s it should have had a more convincing operational record. But in post-Korean-War operations against the Chinese Nationalist Air Force it has not given any great show of might. After the almost purely defensive fighter operations of the Korean War, it was not impressive in intercepting the small-scale leaflet and bomber raids made by Chiang Kai-shek's squadrons based on Formosa against Amoy, Shanghai and Canton. Early in 1955 came the evidence that a tactical air force equipped with MIG 17 fighters and IL 28 twin-jet bombers had been formed and trained. Mixed forces of 50-75 of these machines carried out a series of air attacks on the Tachen Islands, part of the outer defences of Formosa. The attacks were irregular and spasmodic and soon stopped. It would be interesting to know what the crash rate and general aircraft wastage figures were at the time. The brief period of these attacks makes one suspect that the Chinese Communist air regiments were quickly reduced in strength. Nor did the Chinese Communist air force shine against Chiang Kai-shek's air units in the combats of 1958 when the artillery bombardments of Quemoy and Matsu provided once again a flash point for the news of the day. On this occasion, Chinese Nationalist-flown US Supersabres armed with Sidewinder air-to-air guided missiles seemed to have more than held their own with Chinese Communist regiments of jet fighters.

In short, before China can become a large modern self-sufficient air force she has a long haul of a decade or more. But she has made an impressive beginning. Soon her air force of some 3,000 combat aircraft will be all jet and include some 2,000 subsonic MIG 17 and supersonic MIG 19 fighters. Her current tactical bomber forces of IL 28 twin-jet machines comprise some 500 machines. Modernisation with subsonic Soviet-built YAK 25 twin-jet machines is likely soon and these could easily reach Formosan targets. Re-equipment with supersonic Ilyushin twin-engined bombers and faster Mach 2 MIG 21 fighters is likely in the next year or so. More significant is the development of a Sino-Communist air transport and parachute troop arm which at the moment comprises about a division of 10-15,000 parachutists. Chinese women are trained and used both as parachutists and transport aircraft pilots. However, the present equipment of the air transport regiments is obsolescent or obsolete, consisting as it does mainly of LI 2 twin transports of World War II vintage and Ilyushin 14 twin-engine transports both of which will probably be replaced gradually by Soviet jet transport machines, possibly the four-jet turbo-prop Antonov 10 or the four-engined turbo-prop Ilyushin 18. No doubt too some of the present MIL 4 assault helicopters will be supplemented by the large YAK 24 assault helicopters.

The Chinese Communist Air Force is divided into some 80 air regiments on the Soviet pattern. The regiments usually have three squadrons and three regiments usually comprise a division. However, the long-range bomber division created in 1955 following the Soviet despatch of TU 4s to a Peking air base, may have only two air regiments. It will be interesting to see if the USSR provides the Sino-Communist Air Force with any really modern long-range jet bombers in the next year or so, and whether the reports circulating in the past two or three years in Warsaw and elsewhere of the Soviet delivery of a stock of atomic bombs to Communist China are really substantiated.

Whatever the future, it is clear that the Sino-American Defence Pact of 1955, under which the United States committed herself to the defence of Formosa, has acted as a deterrent to the full exericise of Communist Chinese air power during the past five years. China needs breathing space and time to develop her radar early warning system, to strengthen her fighter forces and to modernise her long-range bomber and air transport arms. She is slowly developing to these ends and is being helped not only by the Soviet Union, but by Poland, East Germany and Czechoslovakia. These countries have in the last few years sent China flying instructors, gliders, helicopters, electrical equipment, aluminium and fuel. It is to China's technological credit that only a decade after the Communists came to power her factories turn out most of her ammunition, bombs and aero-engines for trainer aircraft, and that she can assemble modern jet planes. The current decade will bring further technological progress which may surprise us all.

The most important air allies of the USSR in Europe are Poland, Czechoslovakia and Hungary. The East German Air Force has an uncertain political future. A settlement in Germany could neutralise its value as a potentially Soviet ally. If, however, there is no settlement of the German question, then the new Luftstreitkräfte could become the most significant of Russia's European air forces in the Warsaw Treaty alliance.

Largely because of the Hungarian Revolt of October 1956, the Hungarian Air Force is the least modernised of the four and is likely to remain so at least for the next few years. Members of the Hungarian Air Force have the distinction of being the only post-war allies of the USSR who have, so far, carried out operations against forces of the Red Army. A few days after the beginning of the October 1956 Revolt, a handful of pilots and ground staff, dubbing themselves the "National Hungarian Air Force" made a few sorties against Soviet anti-aircraft batteries near Budapest. They also engaged in ground combat with Soviet Army troops and tanks. Their heroism had little effect, politically or militarily, except to reinforce the Soviet Air Force air regiments in Hungary and to slow up the process of modernisation of the Hungarian Air Force which the Soviet authorities temporarily disbanded in the autumn of 1956.

Virtually the only combat experience of the Hungarian Air Force has been its operations with the Fourth Luftwaffe Air Fleet in the Second World War. By the summer of 1943 the German air authorities had equipped most of the squadrons with Junkers bombers and Me 109 fighters. Their air operations in that year on the Ukrainian Front, and in the next two years from Roumanian and Hungarian bases were largely ineffective. The Soviet air armies opposing them were overwhelmingly strong and they were flying modern planes such as the Tu 2 and Ilyushin 10 Sturmoviks and Yak 3 and Lavochkin 7 fighters. When the communist régime was established in Hungary in 1947-48, the USSR equipped the Hungarian air units with obsolescent and obsolete planes such as the Yak 9 piston-engined fighters and ironically some of the Tu 2 and Il 10 Sturmovik ground-attack planes that had been used against the Hungarian Air Force by Soviet air regiments in the 1944-45 air operations. Hungarian air transport units were given obsolete Lisunov 2 twin-engined machines, and the Hungarian Air Force trainer planes, the Po 2 and Ut 2, were of even earlier vintage. In the 1949-50 period, the Kremlin changed its policy and began to equip the Hungarian Air Force with jet planes. At first they received the obsolescent Mig 9 fighter bomber and the Yak 15 fighter. But by 1953, some 650 m.p.h. Mig 15 jet fighters were beginning to equip the tactical fighter air regiments of the Hungarian Air Force and the twin-jet Ilyushin 28 began to replace the Ilyushin piston-engined Sturmovik aircraft. In the same period the more modern Ilyushin 12 twin-engined transport planes began to replace the obsolete Lisunov twin-engined aircraft. Just before the Hungarian Revolution of 1956 there was further modernisation with the subsonic 700-720 m.p.h. Mig 17 and some improvement in Hungarian radar equipment and also the introduction of a few Soviet Mil helicopters. At the same time the operational overall front-line strength rose from some 200 to nearly 300.

Naturally the Hungarian Revolution put the clock back and the air force was temporarily disbanded by Soviet occupational forces. In the past four or five years there has been much rebuilding and retraining under Moscow auspices of the Hungarian Air Force tactical units. There are now probably some 250-300 front-line aircraft most of which are Mig 17 fighters and Il 28 twin-jet light bombers. Some expansion and re-equipment with the supersonic Mig 19 is being carried out and further modernisation with supersonic Yakovlev or

Ilyushin fighter-bombers is likely within the next few years. The transport planes are mainly twin-engined Ilyushin and Antonov 2 biplane machines. There is a division of airborne and parachute troops in being. This suggests that the USSR will probably re-equip the Hungarian air transport arm with one of the three long-range jet planes currently under production in the USSR, but this is not likely to occur before the Czech and Polish Air Forces have been so equipped, for these are clearly more important among the Soviet air allies in Europe.

Hungary's chief value is perhaps as a supplier of bauxite for the Soviet aircraft industry. Hungarian aircraft production is limited. There is some assembly of Soviet-designed jet planes at Gyor and Adyliget and no doubt some further developments will occur in the next few years. But Hungary is not likely to be a vital air ally in the hierarchy of the Warsaw Treaty powers. She is however a useful base for Soviet intermediate-range rocket weapons of the T 2 type. Bases for these weapons have been established near Lake Balaton and there are reports of other Soviet rocket weapons near Szolnok and Debrecen.

Higher up in the Communist air hierarchy in Europe are the air squadrons of Czechoslovakia, Poland and the Soviet Zone of Germany. They constitute important elements in what the Soviet General Staff has called the "water towers" of the Soviet Union, that is its early warning and outer defence system. As with the Hungarian Air Force, training under Soviet auspices has in all three countries been directed at creating a modern tactical air force with a small parachutist arm; but these three European Communist air forces are being fitted with supersonic MIG 19 and possibly MIG 21 planes. The best fighter squadrons are being trained to co-operate with the Soviet PVO strategic defence fighter air arm. It is probable that at the moment Czechoslovakia is the most stable and therefore the most important of these three European air allies. She has very important uranium deposits which have made a vital contribution to the stockpile of Soviet atomic weapons. She has a well-organised modern aircraft industry which makes jet fighters and medium bombers and a useful crop of modern trainer and transport planes. Moreover the past flying record of Czech pilots, whether on duty with the Soviet, French or Royal Air Force in the Second World War, suggests that the Czech Air Force if called to battle would give as good account of itself as equivalent air forces in NATO.

In the immediate post-war years, the Czech Air Force had fairly high training standards and good morale, but there were enormous difficulties in getting suitable modern aircraft for the operational squadrons and the flying training schools. The Czech aircraft industry had been developed by the Germans in the 1939-44 period. Both Heinkel and Messerschmitt built modern aircraft plants on Czech territory. But during the final year of World War II, these and other aircraft factories were devastated by Anglo-American and to some extent Soviet air attack. In the early post-war years the Czech Air Force kept itself going with a curious mixture of obsolescent Yakovlev and Lavochkin fighters and some Ilyushin and Tupolev tactical bombers all of Soviet design; they also had some out-of-date German Arado and Siebel trainer aircraft and a few squadrons of Spitfire and Mosquito aircraft supplied by Britain.

Early in 1948 both the political and air force situations changed violently in Czechoslovakia. The Kremlin's political coup which brought the Communists to power resulted in a major purge in the Czech Air Force. Roughly half of its corps of senior officers were dismissed because they had British wives, or had served in the Royal Air Force or the French Air Force or were otherwise suspected of sympathy with the West. And for the next few years the quality of the Czech Air Force was to be diluted by the loss of men of the calibre of General Karl Janousak, who had been Commander-in-Chief of the Czech Air Force in Britain between 1940 and 1945. Large numbers of Czech airmen, engineers and technicians left their country in this period. The personnel standards of the Czech Air Force in the early 1950s bore the marks of these dilutions. The early years of Soviet indoctrination included the usual compound of Marxian dialectics and the compulsory study of Russian and Soviet history and politics. But there was also a process of modernisation with Soviet jet aircraft, and up-to-date training facilities were provided in the USSR. This led to a temporary period of harsh discipline in the Czech Air Force and relations with Soviet Air Force personnel and staff were far from satisfactory. In the early 1950s the Czechs were given only obsolescent jet planes for their operational training, such as the German twin-jet Me 262 and the Soviet built twin-jet MIG 9 and single-engined jet YAK 15.

But since the 1952-53 period there has been growing harmony with the Soviet air authorities. Virtually all air training has taken place in Czechoslovakia and the Czech Air Force has been equipped with fairly modern Soviet planes. The fighter regiments were soon equipped with MIG 15s, and the tactical bomber regiments with twin-jet IL 28s. By the middle 1950s the Czech Air Force was receiving 720 m.p.h. MIG 17s from USSR factories and was building its own jet aircraft as well. At first the Czech-built MIG 15 jet fighters (called the CS 102) were powered by RD 45 engines from Soviet production resources, but a later version, the S 103 built in Czechoslovakia at factories near Prague and Bratislava, was powered by Czech-built, though Soviet-designed, Klimov turbo-jet engines. Some of these planes were supplied to the Egyptian Air Force in 1955-56, and later others to the Syrian and Indonesian Air Forces. And in mid-1957 the expanding Czech aircraft industry began to retool to make its own version of the supersonic Soviet fighter, the MIG 19. It was by then already making its own twin-jet version of the Soviet IL 28 as well as the Soviet twin-engined transport machine the IL 14, a range of trainer and communications aircraft such as the TOM 8, the L 60, the CD 40 and the ZLIN 26. It was also beginning to turn out small- and medium-sized helicopters of independent design. In addition Czechoslovakia began to develop an electronics industry of its own to supply the Czech Air Force with native-built radar, wireless and radio telephone equipment.

It is clear that the present Czech Air Force is a useful asset to the Warsaw Treaty powers' combined air strength. Politically it is probably more reliable and technically it is at least as advanced as any of the other of the European air allies of the USSR. Its front-line strength is believed to have increased from some 600 to 750-800 aircraft in the past year or so, and may increase further to a strength of nearly 1,000 planes. Nearly all the machines are organised into air regiments on the Soviet pattern, i.e with a strength of 30-40 planes per unit. Most of the air regiments are fighter units equipped with Czech-built machines equivalent to the Soviet MIG 15 and 17, but there is an increasing number of supersonic MIG 19s in the air force and re-equipment with this plane, also perhaps the MIG 21 may become general in the next year or so. The bomber force is purely tactical and equipped almost entirely with Czech-assembled or Czech-built IL 28 twin-jet bombers. There are some half-a-dozen air transport regiments which fly mainly IL 14 twin piston-engined planes, but some re-equipment with four-jet turbo-props., either the Soviet-built IL 18 or ANT 10, is likely within the fairly near future, i.e. before the end of 1962. There is at least one division (and possibly two) of Czech parachutists who would benefit as a fighting force from re-equipment with the new Soviet long-range jet transport planes.

Thus in the last few years the Czech Air Force has been modernised, better trained and fully integrated into the Warsaw Treaty Alliance. It is almost entirely a tactical air force, having no force of strategic long-range bombers and no self-contained strategic air defence force on the lines of the Soviet PVO. But its MIG 19 fighter squadrons supported by the improving quality of the radar early warning defence posts established in Czechoslovakia could make a useful contribution to the outer air defences of the USSR in the event of attack by manned bomber formations. Finally Czechoslovakia provides bases for Soviet intermediate-range rocket missiles in a number of centres near Karlsbad, Reichenberg, Olmütz, Budweis and Javorina, though there is no indication so far of the development of a Czech manned long-range rocket force.

The Polish Air Force development has been similar in many ways to that of the Czech Air Force, but the Poznan riots, the departure of Marshal Rokosovsky and the advent of the Gomulka régime in 1956 has made its political and military adherence to the Soviet régime rather less certain. Officially Poland is a fully fledged and loyal member of the Warsaw Pact. Basically the Polish Air Command is pro-Soviet and Soviet-trained. Since 1943 when the first Polish Communist air squadrons were formed in USSR territory to fight against the Luftwaffe and the Wehrmacht the Commander-in-Chief of the Polish Air Force has been a Soviet citizen or Soviet-trained. From 1943 to 1945 General Polynin was the C.-in-C. and later the Soviet Generals Romeyko and Turkiel.

As with Czechoslovakia there was both contraction and expansion in Polish aircraft production during the Second World War. The Luftwaffe set up both Heinkel bomber and Focke-Wulf fighter plants on Polish territory as well as V 2 rocket-firing centres at Blitzna and Heiderlage. But the Anglo-American bomber offensive of 1944-45 achieved major damage and dislocation against Polish aircraft production. In the past few years aircraft production has been restored and expanded at plants near Warsaw and at Rzeszow, Mielec and at Swidnick near Lublin. The front-line strength of the Polish Air Force is about 750 aircraft and its aircraft industry is probably turning out planes at the rate of 100-150 aircraft per month; but a high proportion of these are trainer and communications aircraft. Polish air regiments are organised on the Soviet pattern and are nearly all tactical in function. The big question is, what has been the effect on the Polish armed forces, including the Air Force, since Spychalski was appointed Minister of Defence after the political upheaval in Poland of the summer and autumn of 1956, when Golmulka came to power. Speaking to senior Polish politicians Golmulka is reported to have said: "Now we will manage ourselves." Certainly Soviet Communist Party interference with the Polish Air Force has been lessened since 1956 and the number of Soviet air personnel in Poland has been reduced. Education and ability are now more important elements in the selection of officers and other entries into the Air Force from the aero-clubs. Both these clubs and gliding schools have been greatly expanded in the last four or five years and are using Polish-built equipment such as the Junak 2 and 3, the Kania, Bies and Kos gliders, some of which are being made under licence in Communist China. The Soviet YAK 12 and 18 trainers are also being built under licence and use is also made in the aero-clubs of the Czech Zlin 26.

But the biggest change since October 1956 has been in the reorganisation of the Polish aircraft industry. A Decree of the Ministry of Heavy Industry issued in October 1957 centred aircraft design and production at Okecie near Warsaw and a new production body called the OKL has grouped together Polish aircraft designers and engineers who are making their own planes independently of the USSR. Soltyk for instance has produced a new jet trainer, the TS 11 Iskra, capable of a speed of over 500 m.p.h. and powered by a Polish-designed jet engine giving a thrust of between 1,200 and 1,300 lb. Misztal and Duleba have produced the MD 12 four-engined transport aircraft at first powered with conventional piston-engined units but later modified and powered by turbo-props, designed by Narkiewicz. Zurakowski is working on two jet-engined helicopters, the BZ 4 and JK 1, which may be equipping Polish Air Force units soon. The Soviet helicopter M 1 is also being built and is designated the SM 1. The Soviet YAK 12 trainer has the title PZL 101 and is being built at a Warsaw factory. The MIG 15 being built at Mielec has the designation LIM 1 and LIM 2. There is also production of the 720 m.p.h. MIG 17. The traditional pre-war PZL title has also been given to a new Polish-designed trainer and communications aircraft, the PZL 102. Finally there are reports that the Polish aircraft industry is doing independent work on guided weapons and atomic-engined aircraft.

The revived Polish aircraft industry is certainly becoming more and more independent of Soviet designs and Soviet supplies. It has had some modest export successes with its helicopters to Syria and Egypt and with its gliders to China. How far and how soon will it become independent of Soviet control and completely national and self-contained? This is a question which can only be answered by the events of the next five or ten years. In 1956, the ten air divisions of jet fighters and light bombers with the division of parachute troops constituted an important element in the strategic air resources of the Kremlin in Europe. But since then the political atmosphere of Poland has altered and the Polish Air Force is acquiring gradually a new sense of national independence. As a member of the Warsaw Treaty Alliance Poland remains in the Communist camp, but as a less stable element from the Kremlin's point of view. In combat the Polish Air Force would display all its traditional skills and bravery. How hard the air squadrons would fight with the Soviet Air Force is surely doubtful.

Meanwhile, Polish youth is becoming more and more air-minded as aero-club facilities expand in all parts of the country, including Warsaw, Katowice, Bielsko, Wroclaw and Gliwice; and the aircraft industry looks forward to an independent future in which it may enjoy an expanding export market.

In an article in the *Sunday Times* during July 1959, Field-Marshal Montgomery said that, "Politically, economically and geographically the German Democratic Republic is Russia's most important satellite." He might have added "aeronautically", for the East German Air Force is potentially the most modern and most powerful of the air allies of the USSR in Europe. It is estimated to have between 300 and 350 jet planes—mainly MIG fighters, both the subsonic MIG 15 and 17 types and some supersonic MIG 19 squadrons—and a growing tactical bomber air force fitted with IL 28 and YAK 25 twin-jet machines. However, there are said to be expansion plans to give it a front-line strength of over 1,000 aircraft in the early 1960s. Until 1955 the Soviet authorities did little to promote military flying in what had been enemy territory and moreover an enemy that had nearly brought about their downfall in 1941. The so-called Volkspolizei ("Civil") Air Force of the previous five years or so had only obsolescent or trainer aircraft of the YAK and PO 2 designs. But in the last five years the strength of the flying personnel of the new East German Luftstreitkräfte has been roughly doubled from 5,000 officers and other ranks to some 10,000.

What is the future rôle of General Heinz Kessler's air force in Soviet air plans? It might be used if a clash with Western Germany arose, so as not to involve Soviet armed forces initially. It is being developed to form part of the outer defences of Marshal Birusov's strategic air defence command. It has no important rôle in Soviet long-range air attack plans except to provide long-range rocket bases near Erfurt, at Peenemunde, on the island of Rügen and in Thuringia. East German pilots until a year or so ago did most of their advanced operational flying training in Czechoslovakia and the USSR. This training on jet fighters began as far back as 1952. But they now do most of it at their own air bases such as Werneuchen and Oranienburg near Berlin, or at bases near Hagenow, Cottbus, Neuruppin, Würzen, Torgau, Wittenburg and Magdeburg.

One of the most interesting recent developments has been the test-flying of a new four-jet turbo-prop. air liner, the B 152, of which the engines are being manufactured at Chemnitz and the air frame at Schkeuditz. This aircraft is designed by Professor Baade. Until the emergence of this plane in 1958, the East German aircraft industry had been making Soviet-designed engines and components and accessories not only for themselves but also for the aircraft industries of Poland, China and Czechoslovakia. One wonders if the B 152 transport plane is not the first sign of East German unwillingness to adhere to complete Soviet standardisation and to conform to Soviet requirements. At a time when the USSR is building three large four-engined turbo-prop. passenger and military transport planes, the TU 114, the IL 18 and the AN 10, it is peculiar that Baade should be launching out on a design of his own. The revival of a strong military air force in the German Democratic Republic could enable the Russians to provide local air pressure in the Berlin area, but as a general air asset in Europe, the new Luftstreitskräfte has limited possibilities. There are incidentally reports that it is being fitted with Mach 2, MIG 21 fighters.

The main importance of the other European air allies of the USSR, Albania, Roumania and Bulgaria, is that they provide valuable dispersal for Soviet aircraft and rocket bases. Albania relies entirely on Soviet help for training and maintaining a small air force of about 50-100 MIG subsonic fighters and a small force of IL 28 tactical bombers. But there are also reported to be Soviet IRBM T 2 bases near Durazzo and Valona. Roumania and Bulgaria each have tactical air forces of some 350 and 150 aircraft respectively, with the Bulgarian Air Force less modern as well as smaller. In addition they are each forming a division of parachute troops. Roumania assembles Soviet-built jet fighters and builds some of its own native designed trainer aircraft. Bulgaria is more important as a dispersal area for Soviet long-range jet bombers which use advanced bases near Kumarica and Dmitrov. Roumania because of her oil production has a more modern fighter force, including a few squadrons of supersonic MIG 19s.

In the major plans of Soviet strategic air attack and defence, the operational value of her air allies in Europe is secondary. They provide part of the early warning system of the USSR but by no means a crucial part, though they are valuable as "water-towers" in the first lines of air defence in the event of an attack from the west or south-west. They have been and are valuable in providing bases for Soviet intermediate-range rocket missiles but this value will be greatly diminished in the next few years when the mass production of Soviet ICBM missiles will take place, unless some form of disarmament agreement materialises. The Warsaw Treaty countries are both a platform and a springboard for Soviet air operations. Their air forces are not major factors in Soviet long-term and long-range operational plans. Their value is that they can be used in local limited wars without involving Soviet armed forces. They can raise stakes without involving a major gamble on the part of the Kremlin, a gamble that the USSR is anxious to avoid.

The last five years have seen a setback to Soviet plans to develop air allies in the Middle East. Soviet MIG fighters and ILYUSHIN tactical bombers have largely replaced the British aircraft which previously equipped the Egyptian and Syrian Air Forces. Soviet flying instruction has replaced British. Soviet radar engineers have established an early warning system in Egypt and Syria, and Soviet constructors have built new air bases in the Arab world. But Soviet plans suffered a setback in 1958-60 when Colonel Nasser took exception to Communist penetration of the United Arab Republic. No longer can the Kremlin plan to use the air forces of Egypt and Syria to act as trouble shooters when they wish, or to provide advanced air bases for operations against U.S. air and naval forces based in the Mediterranean area. No longer are Egypt and Syria reliable potential springboards from which to leap into Africa.

The Soviet authorities have acted with prudence and restraint in the face of their Middle East disappointments of the past two or three years, just as they did in October-November 1956 during the war between Israel and Egypt and the landing of Anglo-French forces in the Suez Canal Zone. Then they broadcast threats to bring in volunteers to help the Egyptians and to bombard London and Paris with rockets. It would have been so easy for a Soviet pilot in Egypt to "direct" a few Ilyushin bombers against British and French planes which were packed wing-tip to wing-tip in Cyprus. But Soviet conduct was a model of prudence and restraint, as it was later when British and American troops landed in Jordan and Lebanon. Soviet caution in avoiding global war is illustrated in perhaps no better context than in her lack of action in Arab territories during the past five years. In 1955 we were inclined to count the United Arab Republic amongst the potential air allies of the USSR. In 1960 it seems as though the USSR is losing as much military ground in the Middle East as Britain did in the earlier post-war period.

CHAPTER 14

SOVIET AIRCRAFT AND ROCKET PRODUCTION

BEHIND the remarkable story of Soviet aircraft and rocket developments of the last decade lies the Soviet aeronautical industry. There is enough information available to show its main developments in the last forty years and to predict probable lines of future development. It would, however, be idle to pretend that there are enough data available to document all the design and production developments in many sections of the industry. This is particularly true of the production of long-range rockets in the 1950s.

The documentation is most plentiful of course in the case of the earlier periods during which the Soviet government relied heavily on foreign technicians and aircraft designs. Just before the Revolution of 1917, the Russian aircraft industry was making between 1,500 and 2,000 planes a year. Tupolev, reviewing the pre-revolutionary aircraft industry of Russia, estimated that it employed 10,000 workers in sixteen aircraft plants. Then came the years of chaos, disruption and shortage of engineers, raw materials and skilled workers. By 1919, only five or six of the sixteen aircraft plants were repairing or producing machines for the new Soviet Air Force. Up to 1925, most of the Soviet planes were of foreign design or supplied by foreign powers. Following the secret agreements between the Red Army and the Reichswehr, Junkers helped the Soviet to establish manufacturing facilities at the Fili plant near Moscow and French aircraft engineers helped to set up aero-engine production facilities in another factory near the capital. Gradually the USSR began to manufacture new aircraft engines, but they were nearly all of foreign design—the French Gnome-Rhône, and Hispano-Suiza and the German BMW. From the United States there was the Wright Cyclone and from England the Napier. Many of the airframes were also of foreign inspiration, but by the mid-1920s TSAGI, the Soviet authority for the design and development of aircraft, had assembled a keen corps of aeronautical designers, such as Polikarpov, Grigorovich, Kalinin and Poutilev. In addition, there was Tupolev, the doyen of Soviet aircraft engineers, and his tough-minded perfectionist pupil, Sergei Vladimirovich Ilyushin, as well as the autogyro and helicopter designers Mil and Bratukhin. By 1922, Tupolev had designed the first new Soviet machine, a single-seater low-wing monoplane, the ANT 1. Within five years or so, the Soviet aircraft industry was turning out native-assembled aircraft at the rate of 500-750 per annum, at about a dozen aero-engine and airframe factories in or near Leningrad, Moscow, Rybinsk, Voronezh, Zaporozhe and Taganrog.

But it was to take more than a decade to bring order out of chaos in the Soviet aircraft industry. Shortages of skilled workers, aluminium and rubber, shoddy finishing of the aircraft products, and lack of adequate machine tools were just some of the factors that slowed up the progress of the Soviet aircraft industry in the early 1920s. In an attempt to reorganise, expand and re-build, the Soviet government introduced a series of Five Year Plans. The first covered the period 1928-32 and called for a production of just over 1,000 aircraft per annum. Of these over half were to be fighters and fighter-bombers, and some 200 tactical reconnaissance machines. But room was found for building some 100 twin-engined so-called heavy TB 1 Tupolev bombers. The rest were trainer and naval aircraft. The most striking omission was, perhaps, the absence of any plan to manufacture any quantity of transport aircraft; but this was a period in aviation development in which passenger aircraft and military transport machines were in their infancy. The possibility of transporting large bodies of troops or vital equipment and supplies by air was not yet part of contemporary military thinking. However in 1930 the Soviet government created a separate department for civil aviation and were on the point of creating the first cadres of airborne troops. The series production of fairly large numbers of civil passenger and military transport machines was also begun.

The Soviet Second Five Year Plan for aircraft production covered the period 1933-37. It was crucial in many ways, for it was the prewar period in which the foundations of the wartime Soviet aircraft industry were laid, and a period in which the rising military power of both Germany and Japan offered a strong air threat to the USSR. The first requirement was expansion and in this period the capacity of the Soviet aircraft industry went up from some 2,000 aircraft a year to over 4,000. A major feature of production policy of this period was the reduction in output of naval aircraft, i.e. floatplanes and flying-boats. The M 14 seaplane fighter of the First Five Year Plan period had no successor. The MR 1 and MR 2 bomber and reconnaissance flying-boats and floatplanes were replaced by the smaller scale production of Beriev's MBR 2 single-engined reconnaissance flying-boat. Another feature was the abandonment of the two-seater fighter category of the First Five Year Plan such as the DI 2 (D stands for two seater and I for fighter in Russian). A further key decision affecting the Second Five Year Plan for aircraft production was made by the Soviet Supreme Economic Council which decided to cut radically the output of small communications light transport aircraft, of which four types had been designed by Kalinin and test-flown during the First Five Year Plan production period.

The expansion of the Second Five Year Plan aimed at increasing and modernising the tactical air forces first, then on a lower priority, at building a force of long-range bomber and transport aircraft. To achieve this an increase both in factory space and in trained aircraft mechanics and technicians was needed. The available evidence of factory expansion is more concrete and detailed than the evidence of expansion and improvement in Soviet aeronautical technical education and training. Many uncertain figures in tens and hundreds of thousands have been quoted for the increase in the number of trained aircraft engineers in the Soviet aircraft industry in the 1933-37 period. But the Soviet aircraft industry during this period certainly expanded east of Moscow and beyond the Urals and so brought reality to the pipe-dreams of Lenin and Trotsky, who in the early 1920s had realised the strategic importance of producing aircraft beyond the range of West European bomber forces. It was in the 1933-37 period that a new string of aircraft plants were built and gradually began to produce aircraft at centres such as Tashkent, Irkutsk, Semenovka, Tomsk, Novosibirsk, Sverdlovsk, Kuresk and Khabarovsk. And at the same time the plants in other areas, such as Kazan, Kharkov, Gorki, Saratov and Leningrad were enlarged. By 1938, the new eastern group of Soviet aircraft factories was making about a third of the total monthly output of some 500 aircraft. During the period of the Second Five Year Plan the number of skilled workers in the aircraft industry was probably more than trebled. There were now some thirty major airframe and aero-engine production centres in the USSR, some of them employing as many as 15,000 workers and there were other smaller assembly centres employing some 2-3,000 workers. The Second Five Year Plan brought the Soviet aircraft industry within a short distance of the numerical output of the other major air powers, such as Germany, Japan, Britain and the U.S.A.

But there was still leeway to make-up in a number of important respects. Soviet synthetic rubber production still had to be heavily supplemented by imports of raw rubber from Malaya, Siam and the Dutch East Indies. The aluminium position was much better and there were reasonably adequate supplies from natural sources in Kandalakska, east of Sverdlovsk, in Western Siberia, parts of the Caucasus and in Soviet Armenia. A little was still imported from Central Europe. Copper had to be imported from both Canada and the U.S.A., though increased supplies were being mined in the Caucasus, Siberia and Urals and other parts of Asiatic Russia. Tin had to be imported from Malaya, Siam and China and the Dutch East Indies, and though there was enough crude petroleum available in the USSR, in fact, a reported annual production of some thirty million tons, one million tons of high octane aviation fuel had still to be imported annually for Soviet Air Force needs.

The standards of machine tools, of skilled technicians and of finished parts compelled the Soviet aircraft industry to produce the simpler types of aircraft with the minimum of radio equipment and other blind-flying aids. This enabled them to achieve high output figures, particulary as they concentrated on making only one or two types of aircraft in each main category. But it also prevented the Soviet authorities to some extent from launching into the large-scale production of such planes as the more sophisticated I 17 fighter machine designed by Polikarpov which was on show in Paris at the international air exhibition of 1936. Its armament of four machine guns and a 20 mm. cannon was remarkably heavy for this period.

Simplicity of design is no longer a compromise policy due to shortages of raw materials and engineering skills in the Soviet aircraft industry of to-day. In the pre-war period it was part of a deliberate production plan designed to help the USSR catch up with the aircraft industries of other powers. This was something they were far from achieving in the 1930s. In most of the main categories of aircraft, the quality of Soviet planes during the Second Five Year Plan period was inferior to those of Britain, Germany or the U.S.A. Polikarpov's reconnaissance and ground attack biplanes were obsolescent, as the air operations in the Spanish Civil War revealed. His R 5 reconnaissance bomber derived from a prototype which went back to 1927. The earlier versions of this machine were powered by the M 17 680 h.p. engine. Although this was replaced by the 750 h.p. AM 34 in the early 1930s, it still gave the machine a top speed of less than 200 m.p.h. There was a ground attack version also used in Spain called the R-Z which soon became obsolete and was replaced by the more modern Ilyushin Sturmovik.

But the biggest technical weakness was in the vital category of single-engined fighters. The Second Five Year Plan produced the I 15, I 153 and I 16 single-engined fighters which were technically inferior to such comtemporaries as the Gloucester Gladiator, the Hurricane and Spitfire or the German fighters, the Heinkel 51 and Messerschmitt 109, especially the last. The same relative technical inferiority is true in the case of the Soviet SB twin-engined medium bombers.

What of the heavier categories of bomber and transport planes? Tupolev had in the 1930s built the TB 3 in large numbers at the Moscow and Kazan factories. This four-engined long-range machine was intended to serve the dual purpose of providing air lift for the growing number of parachutist and airborne brigades in the Soviet armed forces and also to become the first real long-range strategic bomber in use in large numbers in any air force in the middle 1930s. By 1935, over 500 of these machines were in use in Soviet air units. Then came a period of obscurity in which Tupolev disappeared from the aircraft production scene. Some say he was banished to Siberia, partly because of the failure of his new six- and eight-engined bomber-transport machines. Best known of them was the PS 124 eight-engined Maxim Gorky transport plane ("P S" stands for *perevozochny samolot*, i.e. transport aircraft). Powered by eight 900 h.p. engines, this pre-war giant first flew in 1934. It crashed in May 1935 following a collision between two of the obsolescent biplane I 5 fighters which were escorting it. In the same year, Tupolev began building a six-engined transport successor, the ANT 20 (the Maxim Gorky had also been called the ANT 20); but probably only about twenty of these aircraft were built, and as Tupolev faded temporarily from the aircraft production scene, so the construction of heavy multi-engined bomber and transport planes was given a low priority. Lisunov, a younger aircraft designer, was sent to California to visit the Douglas aircraft plant and to arrange for the construction of a Soviet prototype of the DC 2 and DC 3 twin-engine U.S. transport planes. This was to emerge during the Third Five Year Plan period under the Soviet titles of PS 84 or LI 2.

Soviet aircraft production of the Second Five Year Plan was thus a story of simplicity and quantity production rather than of qualitative achievement. It did not bring Soviet planes to the technical levels of her prospective air opponents. However, partly as a result of the Stalinist purges, a new group of fighter aircraft designers were to achieve greater pre-eminence and to close gradually the technical gap which existed between the USSR and other Western air forces in this vital category.

In the period between the beginning of the Third Five Year Plan (1938) and the outbreak of the German-Soviet War in June 1941, a series of new single- and twin-engined aircraft were designed, test-flown and produced which were a great technical advance on the tactical bomber and fighter aircraft produced in the Second Five Year Plan. Alexsievich Lavochkin, of whose design work little had been heard before the Stalinist purges, teamed with co-designers Gorbunov and Gudkov to produce the I 22 fighter, test-flown in the spring of 1939. Like other Soviet planes of this pre-war period, it changed its name. The Soviet designers were now credited in the new aircraft titles. Thus the I 22 fighter became the LAGG 1, designed by Lavochkin, Gorbunov and Gudkov, and the I 24 fighter, designed by Yakolev, became the YAK 1; this was test-flown in the spring of 1940.

In terms of aircraft development, it is perhaps not surprising that Stalin played for time by temporarily coming to terms with Hitler's Third Reich. Stalin's new single-engined fighter aircraft, which were to fight the battle of local air superiority and to provide cover for his tactical bombers supporting the Red Army ground battles, were very much in an interim stage of production. While this was not of course the main reason for the treaty with Hitler, it was probably one of the important ones. And in the field of tactical bombers, both the new IL 2 Sturmovik designed by Ilyushin and the new fast twin-engined PE 2 bomber designed by Petlyakov were both in the prototype test-flying stage in the crucial 1938-39 period. The PE 2 completed its flight trials in 1940 and went into production in that year.

In the 1938-41 period, the cutting back of the production of four-engined bombers and the concentration on output of single-engined fighters and single- and twin-engined tactical bombers enabled the Soviet aircraft industry to more than double its output from 4,000-5,000 planes per annum to some 10,000-12,000 aircraft at the time when Hitler launched his attack in June 1941.

Within a month or so there was major dislocation because of the blitzkrieg advance of the Wehrmacht into western Russia. More than half the Soviet assembly and production centres had to be moved eastwards to safe centres such as Komsomolsk, Semanovka, Irkutsk and Sverdlovsk. The aircraft plants in the Moscow area, those near Leningrad and Kharkov, the Taganrog factory making flying boats and the engine plants at Voronezh and Nikolaev were all transferred eastwards mainly by rail with their technical personnel, jigs and tools. Stalin gave orders that some of the evacuation trains leaving the West were to be loaded with nothing but tools and machinery for the aircraft industry. But despite this Kremlin priority and the Russian enthusiasm and devotion in times of crisis which as Berdyaev has pointed out in *Origins of Russian Communism* has "the property of switching over religious energy to non-religious subjects", Soviet aircraft production suffered inevitably in the latter half of 1941, not only from the evacuation, but also because of the loss of skilled workers to the armed forces. They were replaced partly by unskilled and semi-skilled women. Then there was the uncertain supply of raw materials in the difficult general transport situation of the autumn and winter of 1941. Then, when monthly production dropped to some 500-600 planes a month and the Soviet aircraft factories could no longer replace battle wastage with new planes, the unoccupied Far Eastern air regiments were transferred to the Western Front and obsolescent planes from Soviet flying training schools were flung into front-line service against the Luftwaffe. The flow of British and American fighter planes also offset some of the Soviet output loss in this period; but the main factors in the recovery were the institution of a double shift in the factories and plants east of Moscow and the patriotic fervour of the workers after the evacuation period. Thus the transport aircraft plant and personnel from Moscow making PS 84 Lisunov twin-engined transport planes reached its previous Moscow production output figure within two or three months of being transferred to Tashkent.

By the summer of 1942, the monthly output was again in the region of 1,000 planes as it had been before the German invasion. From then on there was remarkable numerical and qualitative improvement to the end of the German-Soviet War. The 1,000 h.p. M 105 engine was followed by the 1,200 h.p. AM 35 and the 1,300 h.p. AM 38. By 1944, the AM 42, 2,000 h.p., was being made, and the 1,600 h.p. M 82 was also in production. It was later to give 1,750-1,800 h.p. Shvetzov had also designed and produced the 1,650 h.p. ASH 82 engine, which powered the new Lavochkin 5 fighters which gave such a good account of themselves in the battle for Stalingrad and the keen air battles near Kursk in the summer that followed. A later version of the Lavochkin fighter, the LA 7, had a 1,750 h.p. Shvetzov engine which gave it a maximum speed of over 400 m.p.h. The years that Lavochkin had spent at TSAGI under Soviet designers, Tupolev and Kalinin, were now paying dividends, and the technical gap between Soviet native designed fighters and German, British and American fighters was being narrowed. Even in 1942-43 some of the Lavochkin fighters were outspeeding and outfighting some of the Luftwaffe's Me 109s. By 1943-44, the LA 7 was roughly a match for both the German Me 109 and the Focke-Wulf 190 single-engined fighters.

And Yakovlev's single-seater fighters made similar progress in the second half of the Germano-Soviet War during which more than 10,000 were delivered to Soviet front-line squadrons. Like Lavochkin, Yakovlev was a late-comer in the field of Soviet fighter aircraft design and his YAK 1 with its M 105 engine only gave a top speed of some 350-360 m.p.h. But later versions, such as the YAK 3, YAK 7, and YAK 9 marked a series of technical advances in the 1943-45 periods. French pilots of the Normandie-Niemen squadrons found the YAK 3 capable of dealing effectively with both the Me 109 and Focke-Wulf 190 in combat.

In the twin-engine bomber category, there was also both technical progress and expansion in the 1942-45 period. The PE 2, a proven success, was in large-scale production throughout and continued to be turned out even in the post-war period and supplied to the air forces of Poland, Czechoslovakia and Roumania. The IL 4 was replaced by the DB 3 and Tupolev designed a new medium bomber, the Tu 2, which had a performance roughly equal to its opposite number in the Luftwaffe, the Junkers 88 bomber. The backward areas of Soviet aircraft production were the four-engined heavy bomber and transport aircraft categories. Tupolev's TB 7 bomber never went into larger-scale production, nor did its successor the four-engined Pe 8. No attempt was made to build any appreciable numbers of long-range four-engined transport planes. However, the twin-engined Li 2 was built at an estimated monthly average rate of between 50 and 75 in the last year or so of the war.

By 1944, the successes of the Red Army counter-attacks enabled aircraft plants and their personnel to return to the factories in the West. In that year, a total production of 40,000 planes was reached and this, according to a Stalin statement, was roughly the rate of output at the end of the Second World War. It is rather less than half the peak U.S. wartime output, though as so many American planes were of the four-engine type, the numerical comparison is misleading. It is equally misleading to compare the figures with British aircraft production totals, which reached some two-thirds of the Soviet figure. A fairer comparison can be made with German aircraft production, which reached a 1944 peak figure roughly equivalent to that of the Soviet total, or the Japanese output figures which were about two-thirds of the Soviet total.

To the Soviet output figures of World War II must be added more than 15,000 British and American planes received during the last three years of the War under Lend Lease. The question might well be asked: "Did the Soviet Air Force need these machines in the 1944-45 period?" True she was expanding to a strength of some twenty air armies with more than 600 air regiments. But with the enormous numerical preponderance enjoyed over the Luftwaffe which had to meet the needs of a three-front air war, it is at least arguable that U.S. planes were more needed in the final air battles in the Pacific than they were on the Germano-Soviet War fronts. No doubt for good political reasons the cessation or reduction of U.S. Air Lend Lease to the Soviet Union was not feasible.

It is of course a more difficult task to give an accurate assessment of Soviet aircraft production figures since the War, and from the available evidence an even more difficult task to give reliable figures for current rocket production output.

The basis of the post-war production output is reasonably well documented. It is known that in the 1944-45 period, monthly production ranged from 2,500 to 3,500 machines. Colonel Tokaev, who had access to Stalin and to Soviet Air Force documents has stated that the 1944 production output total was about 43,000 aircraft. But he does not make it absolutely clear if this is for newly built aircraft only or if it includes a proportion of repaired aircraft. Two things are however certain. The first is that while in the five years after the war—U.S. aircraft production fell to a figure of less than 250 planes a month and the figure of forty-eight air groups for the U.S. Air Force was the target operational strength for this period, there is no evidence of any comparable cut in Soviet aircraft production or strength. Secondly, the Soviet aircraft industry acquired the facilities of some four-fifths of the German aircraft industry and went on to expand output facilities and to modernise its equipment. Thus for example the Heinkel aircraft plants at Rostock despatched men and machinery to Kiselevsk in 1949 to establish a new aircraft production centre; and two new large engine plants were reported to be built underground at Kuibishev, east of Moscow, in 1948-50; these were turning out jet engines in 1951. The large airframe plant at Gorki was expanded and Molotov became a new major centre for the production of engines and airframes.

The increase in machine tool production, the inheritance from the Luftwaffe of large hydraulic presses (the Soviet have since built many bigger and better ones of their own design), the concentration on relatively few types of aircraft, the relative simplicity of design, the extension of mechanisation—all these factors have contributed to the extraordinary progress made in Soviet aircraft production in the post-war period. In the early post-war years, it is true that the USSR again relied heavily on foreign designs for a number of their aircraft and aero-engines. The Tu 4 strategic four-engined bomber was a reconstruction by Tupolev of the US B 29 Superfortress, the Mig 9 jet fighter and fighter bomber, like the Yak 15 fighter, was powered by German designed jet engines and the Yak 23 and Mig 15 fighters by British designed Rolls-Royce jet engines. Thus the Fourth Five Year Plan for aircraft production of 1946-50 was similar to the First Five Year Plan of 1928-32 in its heavy reliance on foreign inspiration.

Although many thousand German aircraft technicians went on to work in the Soviet aircraft industry, throughout the 1950s their influence became less marked. To-day the Soviet aircraft and rocket industry may be proud of its extraordinary native achievements during the past ten years. In this period it has succeeded in closing the technical gap between itself and the West. Indeed in the fields of helicopter development and long-range rockets, the Soviet technicians and industrialists have established a lead.

The broad resources of the Soviet aircraft industry are estimated to be as follows: it employs between one-half million and three-quarters million men and women in some 400 aircraft, aero-engine and aircraft equipment factories and it probably has the capacity to turn out more than 2,000 aircraft a month when producing mainly single-and twin-engined planes, and perhaps rather less than 1,500 a month when the output emphasis is on four-engined bomber and transport aircraft as it has tended to be in the last five years. Soviet aircraft production potential has been underestimated in some official Western sources and no doubt we are all in danger of overestimating its rocket production potential. Mr. Louis Johnston, when U.S. Defence Secretary, put Soviet 1950 aircraft production at 600 a month and in July 1951, Mr. Emanuel Shinwell, the British Defence Minister, put the figure of 800-900 a month. Unofficial sources about this period pitched the figures much higher. *Interavia*, an air journal of international repute, gave a 1949 estimate of over 2,000 planes a month and Colonel Gregori A. Tokaev in his *Stalin Wants War* put the figure, including trainer aircraft, at over 3,000 a month. It seemed at the time that some of the unofficial estimates were nearer the mark than the official ones; the main reason for saying this is that at the time Soviet air strength was variously estimated at 17,000-19,000 aircraft and in addition it had to provide thousands of planes each year for its allies in Europe. It could scarcely hope to supply aircraft for such commitments from a factory output of less than a thousand aircraft per month.

Broadly speaking, the USSR has the same total aircraft production commitment to-day as she had ten years ago, with the important addition of having to meet the expansion in four-engined jet bomber and transport output and also that in helicopters. Whereas in 1950 Soviet output of planes in these last two categories was perhaps of the order of 50-75 planes a month it has probably increased to well over 200 a month and including all types of light helicopter and transport machines to perhaps over 300 a month.

Soviet Aircraft and Rocket Production

Never before had the Soviet aircraft industry turned out such a range of transport planes, from the small AN 2 biplane of which over 2,000 have probably been built in the last five years or so to the large four-jet engined turbo-prop. IL 18, AN 10 and TU 114, of which perhaps no more than a dozen a month are being built, though expansion in output is inevitable.

Since 1951, when it was estimated that less than one-fifth of Soviet fighters were jet-engined, the Soviet aircraft industry has provided aircraft for an all-jet fighter force which, including tactical and strategic defence fighter regiments totals over 10,000 operational aircraft. Add to this the requirement of making operational training jet fighters of the MIG 15, 17 and 19 types and the supply requirements of Communist Allied Air Forces, and one comes to the conclusion that the USSR is probably turning out 750-1,000 fighter aircraft a month and perhaps more. Moreover in the 1950s the performance of Soviet jet fighters kept pace with those in Britain, U.S.A. and France and the radar and armament have become increasingly sophisticated. By 1953, the subsonic MIG 17 with a top speed of over 700 m.p.h. was in service. In the next year or so radar-equipped MIG 17 fighters were part of the night and all weather fighter air regiments of the PVO. By 1955, the MIG 19, the first Soviet supersonic fighter was in large-scale production and coming into operational use equipped with the first Soviet guided air-to-air missiles, the M 100. This was followed within a year or so by an infra-red guided air-to-air missile, probably called the T 8. By 1958, the MIG 21, capable of flying in level flight at twice the speed of sound was in production. In the last year or so, an even faster fighter aircraft, the MIG 31, capable of flying at roughly Mach 2·5, has been reported in squadron service. In addition, at least two types of supersonic delta-wing fighters are reported to have been in production in the last three or four years designed by Pavel Sukhoi.

According to the latest information available, the Soviet Air Force has a force of some 5,000 tactical bombers, mainly twin-engined. It is thought that to maintain a front-line strength of this order, build special operational trainer versions, such as the IL 28U, and supply many hundreds of these planes per annum to the air allies of the USSR, a monthly production of between 350 and 450 of these machines is probably needed. The mainstay of the tactical bomber force is the Ilyushin 28, of which many thousands must surely have been built since 1950, for in addition to supplying a Soviet Air Force front-line strength of over 2,500 of these machines in the middle 1950s, many hundreds of Soviet-built IL 28s have been in service as the standard tactical bomber in the air forces of Poland, China, Egypt, Roumania and Czechoslovakia as well as the naval air forces of the USSR. Since 1957 a supersonic development of the IL 28 has been coming into service, believed to have the designation IL 40 (some sources say Ilyushin 140). Another ground-attack light bomber, which has in the last year or so been developed into the supersonic class, is the YAK 25 which has also seen squadron service in a night-fighter version since 1955. Indeed it is used mainly as a twin-engined night fighter in the PVO defence command. The supersonic version of the YAK 25 is probably the YAK 42 which is thought to have been in squadron service for the past two or three years. Currently the supersonic versions of the YAK and ILYUSHIN tactical bombers are probably being made at a monthly rate of 20-25 each. It is likely that production of the IL 28 twin-jet bomber will be stopped very shortly and also that of its contemporary, the TUPOLEV 14, twin-jet naval torpedo-carrying aircraft which has been in production at the rate of probably 10-15 aircraft per month for the last four or five years.

The production of heavy and medium jet bombers is believed to be of the order of 25-50 aircraft per month supporting a front-line strength of some 1,500 aircraft. Up to the period 1954-55, there were no medium or heavy jet bombers in large-scale production. In that period the TU 16 twin-jet bomber, roughly equivalent to the US B 47, the four-jet MYASISCHEV "Bison" bomber, the Soviet equivalent of the U.S. B 52 but with a shorter range and finally the four-jet turbo-prop. driven TU 20 "Bear" bomber came into general service and into series production. It is likely that this rate of monthly production of medium- and long-range jet bombers will continue despite Soviet progress in long-range missiles. In the last year or so the series production of a supersonic delta-wing strategic bomber with an intercontinental range has been reported several times. The modernisation of some of the DA long-range squadrons with jet bombers is barely completed and new machines of this kind are needed for operational bomber training and for the re-equipping of those naval air squadrons which still have TU 4 pistoned-engined machines for long-range reconnaissance.

One can do little more than hazard a guess at the extent of Soviet missile and long-range rocket production, though there is plenty of evidence that it has been ahead of U.S. production and development in most ways throughout the 1950s. It appears that chronologically the output of defensive ground-to-air missiles, such as the T 5, T 6 and T 7, and tactical attack missiles, such as the T 1, preceded the output of T 2 intermediate-range missile by several years and that the key weapon, the intercontinental ballistic missile, went into series production some time in 1958. Reasonably reliable reports from German sources have shown that since 1953 the T 5 four-stage solid-fuel anti-aircraft missile has been in service with PVO anti-aircraft batteries near Moscow, Leningrad, Stalingrad and at key atomic weapons centres such as Kuibischev and Ulan-Ude. Production of ground-to-air defensive weapons has been reported for nearly ten years at Tomsk, Kiev, Irkutsk, Badakshan and at a factory north-west of Leningrad. In all, the Soviet missile industry has to provide anti-aircraft weapons for several hundred PVO batteries, perhaps over 1,000. On the basis of this, it is thought that the USSR has been producing anti-aircraft ground-to-air guided weapons of the T 5, T 6,and T 7 types at an annual rate of perhaps 500 a year.

German sources have given figures of 2-3,000 missiles a year for the output of the T 1 missile, the Soviet development of the German V 2; large-scale production of these weapons was reported as far back as 1950. In the next few years there was evidence of the manufacture of T 1 tactical missiles with a range of up to 400 miles in a number of places in the Soviet Zone of Germany such as Nordhausen, Zwickau, Leipzig and Erfurt, as well as in the USSR at rocket factories near Tomsk, Irkutsk, Leningrad and Kiev, also in unlocated factories in Turkestan and Siberia. But it is unlikely that the figure of 2-3,000 T 1 missiles per annum, if originally accurate, has been maintained, as so much of the manpower and materials of the T 1 went to develop the subsequent IRBM T 2 rocket which is believed to have been in small-scale production by the 1955-56 period. What does seem reasonably certain is that by 1955 the USSR had successfully tested a two-stage rocket capable of carrying an atomic warhead with a range of over 1,000 miles and possibly as much as 1,500-2,000 miles. Although some reports said that these missiles were hitting their targets to within 1,500 feet, the subsequent tracking by radar of a large number of test-firings between 1955 and 1957 suggests that the USSR was then experiencing the same instability from which liquid-fuel multi-stage rockets also suffered in their tests in the USA in 1957-58. Subsequent reports that the USSR were making a second type of IRBM, the T 4, with a better guidance system and with the means of being fired from a mobile ramp, suggest that the T 2 may be phased out of production soon. Production of these IRBM missiles has been reported from Vorkuta, Tiksi, the Moscow area and Khabarovsk. Although there was more than one report from German sources that as many as 20,000 Soviet IRBM's had been produced by the summer of 1958, one feels that this figure was grossly overestimated. An annual production of a few hundred a year of the T 2 and T 4 types of IRBM seems more realistic and since the summer of 1958, there has surely been concentration on producing the maximum number of T 3 and T 3A intercontinental ballistic missiles at the expense of IRBM output.

The T 3 is a three-stage liquid-fuel rocket with an intercontinental range of 5,000 miles and the ability to carry a thermo-nuclear warhead. It was first successfully fired in August 1957, though component testing is reported to have been begun in 1955. Production in the last two years may have totalled several hundred; a figure of between 300 and 600 has been given by various sources. However, in the last year or so, an even longer-ranged and more powerful Soviet ICBM, the T 3A, has been reported under development and successful test-firings of it were reported in 1959. The T 3A is stated to have a range of over 6,000 miles and to be powered by rocket motors, similar to those which launched Sputniks 2 and 3 and which provide a take-off thrust of some 240 tons. It is thought that production of the T 3A is so far only on a very small scale.

The above long-range offensive rockets are all of the liquid-fuel type. But it is reported that the USSR is also manufacturing the Komet 1 solid-fuel rocket, equivalent to the T 1 liquid-fuel missile, and a Komet 2 solid-fuel rocket, equivalent to the T 2 IRBM rocket. It would also be surprising if an ICBM solid-fuel rocket were not being developed as a USSR equivalent to the US Minuteman. Thus the overall long-range rocket production policy would seem to be in a state of transition as it is in the United States.

It would certainly be surprising if the USSR were not having difficulties in storing the liquid fuels for T 3 and T 3A ICBM rockets at the missile sites at both Novazemlya and Severnaya Zemlya, as well as at other long-range rocket sites along the Arctic Ocean between the Kola Peninsula and the Nordvick Cape. Another element of uncertainty must surely be in the guidance mechanisms. Although the USSR claimed to have solved this problem in the summer of 1957, and the successful firing of the Lunik and other earth satellites has revealed a high degree of control, there is reason to believe that the USSR cannot yet produce the inertial guidance in her ICBM rockets to the degree of accuracy which would make these weapons a guaranteed deterrent against U.S. air and hardened rocket bases. Some German sources as recently as 1959 were talking of occasional errors of plus or minus ten miles in even "successful" Soviet rocket launchings.

Reports of Soviet rocket production have come from every region of the USSR: in the west from Riga, Kiev, Sinolevsk and Chernyakovsk, in the far east from Komsomolsk and Vladivostok. Other reports of long-range rocket manufacture have come from Stalingrad, Irkutsk, Tomsk, Novosibirsk and also places near the Caspian Sea. It would seem that the USSR has dispersed her rocket production as widely as she has her aircraft production. It would also seem that the Soviet missile industry has not in any way interfered seriously with the scale of the aircraft industry. This is of course one of the many instances of the great increase of Soviet technological resources in the last decade. The USSR claimed as early as 1951 that their rocket engines could meet the needs of interplanetary flight; but in the last decade there has been no evidence that any major aircraft plant has been shut down because of the successes of Soviet missile production. This is at once a sign of growing technological strength and a pointer to the fact that manned fighters and bombers have still an important part to play in the foreseeable future of Soviet air power.

CHAPTER 15

ORGANISATION AND REORGANISATION

DESPITE the enormous growth of Soviet air power in the last two decades the government of the USSR has never found it appropriate to have a separate Air Ministry. This is in one sense not surprising, for the basic structure of the Soviet armed forces consists of a single powerful unified Ministry of Defence subordinated to the political control of the Communist Party much more decisively than is ever the case in Western countries. Marshal Zhukov said about a year before he was dismissed from the post of Soviet Defence Minister: "We are proceeding from the fact that victory in future war will be achieved only by the combined efforts of all arms of the armed forces and by their co-ordinated employment in war."

Clearly this must continue to be increasingly the Soviet view as rockets launched from submarines of the USSR Navy gradually take the place of the Soviet ADD long-range air forces, and as these air forces themselves are partly redirected to long-range co-operation with Soviet submarines in the next year or so. As for Soviet tactical air power, which is currently the strongest arm numerically, this must of necessity be even more closely linked to the infantry ground force administration and to that of the tank and artillery arms. The same applies to the Soviet VDV airborne forces.

Perhaps the Soviet air arm which is likely to become most prominent and self-contained is the Air Defence Force, PVO, at present commanded by Marshal Biryusov. Like his predecessors in this post, even in the period when fighter aircraft were the chief instrument of air defence, Biryusov is an artillery specialist. Among the dozen or so military leaders of the USSR who hold the highest military rank of Marshal of the Soviet Union there are no airmen. It would almost seem that no Soviet Air Force officer can attain this rank. The highest eminence to be reached so far is that of Chief Marshal of Aviation, a rank attained by such men as Golovanov,

Novikov and Zigharev, none of whom at present holds an active military air assignment.

Nevertheless as Soviet military concepts have broadened in the post-war period, the Soviet air establishment in the Defence Ministry has expanded and Soviet air leaders have become more prominent and to some extent more independent of army control; but they have never reached the degree of prominence or independence achieved by the chiefs of the Royal Air Force and the United States Air Force since the War. Because of the growing prominence of rocket weapons there is now little likelihood that they will do so in the future. It would for instance be difficult to imagine the USSR appointing an air general or a marshal of aviation to command the Warsaw Treaty forces, as has been done in the NATO forces in Europe by the appointment of General Norstad. It is most unlikely that any Soviet Air Force commander would follow Marshals Zhukov and Malinovsky as Defence Minister unless like Marshals Zhigharev and Biryusov they had a strong army or political background.

Operationally and administratively the Soviet Air Force is less unified than it was during the Second World War, when almost its entire resources were devoted to supporting the Soviet ground forces in tactical air operations. It would however be hard to imagine the present long-range bomber command, the DA forces commanded by Marshal of Aviation Sudetz, reverting to a tactical air rôle under ground forces' aegis, as the Soviet long-range air command did at the end of 1944. When the DA long-range bomber air force was re-established in 1946, the Soviet intention was to build up 3 or 4 long-range bomber air armies each of up to 500 bombers, and to give the command the status of a chief administration subordinated directly to the Defence Minister. This plan has now in fact been carried out. Marshal Sudetz is directly subordinate to the Defence Minister, Marshal Malinovsky, and until recently might have been expected to exercise a high degree of control over the operations of his long-range bomber air regiments. But there clearly must be a current re-appraisal of all forms of strategic long-range striking power in Soviet General Staff thinking and planning. It is believed that Marshal Zhigarev hoped a few years ago to achieve a combined long-range bomber and rocket command within the framework of the Soviet Air Force under the operational control of Marshal of Aviation Sudetz, thus bringing the Soviet Air Force in line with the strategic attack forces of Britain and the U.S.A. Some sources think that Zhigarev's failure to achieve this was the cause of his replacement by Marshal Vershinin as Commander-in-Chief of the Soviet Air Force early in 1957. Certainly it would be logical for the Soviet defence forces to be forming a combined organisation which is a strategic striking force making use of land-based and submarine-launched missiles and of the Soviet long-range bomber force; but it is far more likely that it will be headed by a commander with an army, probably artillery and rocket, background than by someone like Marshal Sudetz with his air force background of tactical and strategic bombing. On the whole therefore the Soviet DA long-range bomber force may become less independent operationally in the next few years than it has been in the 1950s and also reduced in strength. It may become part of a new inter-service strategic attack command headed by someone like Marshal Moskalenko, who is said to be politically well thought of by Khrushchev and who has a good artillery and military rocket background.

Another big organisational change in the Soviet air forces since the end of the Second World War has been in the Strategic Air Defence Forces. In accordance with its increasing importance it has been given the status of a major operational command and a chief administration in the Defence Ministry. This means that in the overall hierarchy of the Soviet Armed Forces, Marshal Biryusov's PVO Air Defence Forces occupy a position of rough equality with the naval and ground forces and with the Soviet Air Force itself. Indeed in the local Air Defence districts into which the PVO command is organised, it takes priority over and can operationally be independent of the local military districts of Soviet ground forces and of the local political district administrations of the USSR. Strictly speaking the PVO Commander-in-Chief is not an air force commander and is likely to be less and less a commander of fighter air forces in the coming decade. He has control of the strategic fighter forces, the radar and other warning systems, the defensive ground-to-air guided missiles and of a declining force of conventional but radar-guided anti-aircraft guns. His unified command is likely to be even more important in the future than in the past but he is likely to rely less on his fighter air forces. In this connection it is interesting to note that General Klimov, who heads the fighter units of the PVO command, is not only deputy commander to Marshal Biryusov the

PVO Commander-in-Chief but is also deputy to Marshal of Aviation Vershinin, Commander-in-Chief of the Soviet Air Force. This may lend support to the view expressed in the chapter on Soviet tactical aviation that in time of war part of the fighter forces of the PVO command might be transferred to tactical air armies to secure local air superiority over the battle front.

It may be worth emphasising that even in the 1940s and the early 1950s when intercepting fighters were the chief means of active operational strategic air defence, the Commander-in-Chief of the PVO forces was not an air force general but invariably an artillery general. This once again shows how reluctant the Soviet defence authorities are to assign high military posts to Air Force generals. This may be part of the reason why Soviet Air Force purges and retirements in the immediate post-war period were apparently on a larger scale than those in the Soviet Army. Chief Marshal of Aviation Novikov was prematurely retired (some say temporarily imprisoned) after a fine record of command in the Second World War. So were Marshals of Aviation Falaleev, Vorozheikin and Astakhov, also Air Generals Repin and Grendal, respectively head of the Soviet engineering and intelligence services. By making the chiefs of the Soviet Air Force realise that they were readily expendable Stalin was able to limit their ambitions. And so Marshal of Aviation Vershinin who had replaced Marshal of Aviation Novikov as Commander-in-Chief of the Soviet Air Force was himself replaced by Zhigarev in 1949. But in 1957 there was again a change and Zhigarev was replaced by his precedessor Vershinin.

The airborne forces have also achieved greater importance in the Defence Ministry hierarchy in the post-war period. Lieutenant-General Margelov, Commander-in-Chief of the Soviet Airborne Troop Command, heads a special administration, holds the position of a deputy minister of defence and in the hierarchy of Soviet military establishment is equivalent to the head of the Soviet tank arm in the Defence Ministry though junior to him in rank. The Soviet Airborne Troop Command also has its own large force of air transport regiments which since 1950 have been under the operational direction of Marshal of Aviation Skripko. In my view there is a tendency to underrate the current strength of the Soviet military air transport arm and of the air lift potential of the Soviet airborne troop command. A figure as low as 500 transport aircraft has been quoted in more than one appreciation. I am inclined to the view that this total should be at least doubled and that an increasing proportion of the military air transport regiments will soon be fitted with four-jet turbo-prop aircraft such as the Antonov 10 and Ilyushin 18. One wonders therefore if the immediate two divisional air lift currently attributed to the Soviet Airborne Troop Command is not in fact a conservative estimate. Of course it all depends on where the troops have to be lifted from and to which destination. No doubt in a short tactical lift over the battle area in which large-scale use of Soviet helicopters could be made a lift of three and possibly four divisions might be possible. If, however, Soviet airborne forces in the Moscow area had to be transported to a sudden flashpoint, for instance on the Persian border, the large-scale use of helicopters would be impractical. In assessing Soviet air lift potential for such long-range emergency transport operations one needs to allow for the fact that a large number of long-range turbo-prop and turbo-jet air liners could be requisitioned from the civil air line resources of Aeroflot, just as Germany used the air transport resources of Lufthansa to air-ferry troops, supplies and equipment to Norway in the spring of 1940 and to the Balkans a year later.

There can be little doubt that in any future war Soviet civil aviation would be largely merged into the military air forces and would thus supplement the air lift potential not only of the Soviet airborne forces but also of the tactical and long-range bomber and defensive fighter or rocket forces. The Soviet General Staff certainly learnt the importance of air transport in the Second World War, and despite the delay in building up a modern long-range air transport arm to meet their strategic needs in the 1940s and early 1950s, we may well find that in the 1960s the expansion of air transport resources both military and civil will be the outstanding Soviet air development. At the moment the bulk of their military air transport resources is suitable for employment only in tactical rôles in support of ground forces. But with the announced Soviet intention of expanding in the 1960s the number of military and civil long-range four-engined jet transport aircraft giving the USSR a strategic air lift potential with intercontinental ranges, it may well be that some new combined air transport command will be set up under the Soviet Defence Ministry. One recent indication of the growing importance of air transport operations was the transfer of Colonel-General of Aviation Loginov to Aeroflot from the post of deputy commander to Marshal of Aviation Vershinin, and Commander-in-Chief of the "Frontal Aviation", that is, the tactical fighter, bomber and reconnaissance units of the Soviet Air Force.

In the Defence Ministry hierarchy there is no separate special administration for the tactical air units of the Soviet Air Force. At that level they are part of the administration of Marshal of Aviation Vershinin's overall departmental responsibilities. Operationally they are subordinated to local military districts in peacetime, and in wartime they would be assigned to an Army Group on a particular front. The military districts are geographical commands and there are eighteen of them at present. The theory is that each of them is assigned a tactical air army from the overall resources of Soviet "frontal aviation". At the end of the war, when the Soviet had as many as eighteen tactical air armies, the theory probably worked. As the present total is believed to be only about twelve to fourteen Frontal Air Armies, this means that one or two military districts in the Soviet hinterland are probably without a tactical air army. All the military districts are directly subordinated to the Ministry of Defence and until the mid-1950s most of them were commanded by Marshals of the Soviet Army. They are now all commanded by generals who held high field commands in the Red Army during the last war. As far as is known, no air force general has been in command of a military district. In them the senior air officer holds the rank of a Deputy Commander in parallel with the deputy commander for tank forces, artillery forces and for political affairs. The Soviet tactical air armies consist of a varying number of fighter, fighter-bomber and light- or medium-bomber divisions each of which is subdivided into regiments with an operational strength of 30-40 aircraft. The overall strength of tactical air armies varies considerably, probably as much as from 500 to 1,500 aircraft. Usually those on the Western and Far Eastern peripheries of the USSR are much stronger than those in Central Russia as they are kept on a wartime footing at a high degree of readiness and maximum operational strength.

Incidentally the ground commander of the Military District also has some administrative control over other Soviet Air Force units such as the long-range strategic bomber force (DA) and the strategic air defence forces which form part of the PVO. But in the case of the tactical air forces, he has also a measure of control over operational training and their use in military manœuvres in peacetime, just as in wartime the frontal air armies would be operationally controlled by the commander of the ground forces on the individual fronts. Under war-time conditions it is likely that up to two or three tactical air armies would be assigned to an individual army group to provide fighter cover for the ground forces, fighter escort for the transport aircraft carrying the airborne forces, tactical reconnaissance, interdiction bombing of enemy communications and air bases and close fighter-bomber support for tank forces. The strong grip of the army on Soviet tactical air forces is illustrated by the fact that in official documents they are sometimes called the "Air Forces of the Soviet Army" and sometimes the "Air Forces of the Soviet Armed Forces". At all events it is clear that although responsible for basic training, equipment and tactics, the Soviet Air Force High Command has frequently to bow to the superior position of the Army in the USSR when long-term questions on the organisation and employment of the tactical air forces are raised in the Ministry of Defence. It is doubtful for instance if the air force would have the major or final say in a decision to take away air units, i.e. regiments or divisions, from one air army and reallocate them to strengthen another air army on another front. The job of the Soviet Air Force is to create the tactical air forces and bring them to the pitch when they are ready to operate. After that Army commanders largely take over; though the detailed records of World War II show that within a tactical army the Air Force Commander can exercise initiative and help make decisions on which targets are to be attacked and in what strength. Certainly at the lower echelons of air division and air regiment the air force commanders work directly under the commander of the air army, so that when it comes to the day-to-day business of doing the job considerable latitude would be given to Soviet Air Force tactical air commanders.

While the conventional peacetime conception of the operational organisation of Soviet tactical air armies is based usually on a centralised command, in which an army group and one or more air armies work together under a single direction, it is likely that there will be occasions—particularly to meet the needs of atomic warfare—on which this broad scheme of things will be broken down into smaller elements. The Soviet General Staff must surely be planning other forms of tactical air support for which the conventional organisation is not suitable. Thus special groups of fighter aircraft, consisting perhaps of air regiments from both the tactical frontal air forces and the PVO strategic defence command will have to be combined to protect and cover large-scale helicopter and air transport operations, either of crucial supplies or of airborne troops, and similar concentrations of fighter units will be needed to cover crucial tactical air reconnaissance or to ensure the protection of tactical atomic rockets and artillery. There was a time when tactical air forces were primarily a substitute for a long-range artillery, but modern tactical rockets have now all the range and more than the destructive power of the fighter bombers and Sturmoviks of the Second World War.

But there is one important aspect of modern ground warfare which may compel a radical change in the organisation of tactical air power in the Soviet armed forces. One refers of course to the large-scale use of tactical atomic weapons by the enemy. This might well make the conception of an Army Group and the concentration of large numbers of ground divisions and tactical air regiments entirely impractical. The basic conception of a continuous concentrated front supported by the concentrated air power of two or three frontal air armies is a notion which could become obsolescent in the next few years. The uncertainty of communications and supply under atomic attack may force the Soviet General Staff to break down their tactical air forces into smaller self-contained units of perhaps one or two air divisions each with their own allocation of helicopters, transport aircraft and local supply facilities. In such circumstances a greater measure of autonomy and initiative may devolve on Soviet tactical air forces, if only because the links with the main army command will be tenuous under atomic combat conditions. In World War II the local air commander of a regiment could of his own volition attack a mobile target of opportunity or attack an alternative target if, for instance, the selected airfield was not occupied or if bad weather made it impossible to locate the bridge, railway or other objective originally selected. But in the chaos of large-scale tactical atomic warfare, it will be necessary for the air force tactical air commander to exercise greater and more frequent initiative, basing many attacks on his own immediate reconnaissance without consulting the army authorities, simply through lack of time or assured contact with them. The free-lance bomber reconnaissance operation may in the future become a major aspect of Soviet tactical air power just as free-lance fighter operations did in certain phases of the Second World War.

The Soviet naval air forces have always had an element of duality both in their organisation and operations. Colonel-General Preobrazhensky is Commander-in-Chief of Soviet Naval Aviation. His operational naval air divisions are subordinate to the commanders of the Baltic, Northern, Black Sea, North Pacific and Pacific Fleets. These Fleets are themselves controlled by the Commander-in-Chief of the Soviet Navy, Admiral Gorshkov, who is one of a dozen deputy ministers at the Defence Ministry, including Marshal of Aviation Vershinin himself. The Commander-in-Chief of the Soviet Navy is likely in practice to exercise operational control over those naval air squadrons giving direct naval support to the fleet commanders by oversea reconnaissance, anti-submarine attacks and minelaying and torpedo-bomber operations. But there is a strong shore-based element of Soviet naval air power which consists of fighters liable to be used in defence of port areas and light and medium bombers which may be used in support of coastal land operations of little interest to the Navy. In such cases part of the operational control of the Soviet naval air units would probably pass temporarily to the local PVO or tactical air force commander. No one is more fully aware of this possibility than Colonel-General Preobrazhensky himself, for he led a twin-engined Soviet naval air squadron of the Baltic Fleet Air Arm on a raid on Berlin in August 1941 on what could scarcely be regarded as a naval air occasion.

But the organisation and control of Soviet naval air operations will call for some rethinking in the 1960s when the numbers of nuclear-powered Soviet submarines capable of firing long-range missiles will surely grow. With this growth, there will be a need for more long-range aircraft to co-operate with these submarines in reconnaissance, communication and perhaps air supply rôles. In the chapter on naval aviation the possible diversion of long-range jet bombers to co-operation with submarines has already been considered. They may in addition to the above rôles also become part of the general Soviet anti-submarines' attack forces. Finally Soviet submarine and naval air forces may in the future form part of an overall strategic bomber and rocket assault command in the Soviet defence forces.

It is clear that the organisation of air forces and air power in the USSR Defence Ministry prevents the Soviet Air Force from achieving political parity with the Soviet Army and Navy as the Air Force has done in the United States. It is of course impossible to make useful comparisons between air forces on the bases of organisation and administration alone. One cannot believe that staffing procedure and organisation is even a remote guide to the practical expression of doctrine. One is reminded for instance that in World War II the United States Army Air Forces, in the khaki uniform of the ground forces, produced the most successful strategic air operations of the Second World War; while the Luftwaffe which enjoyed independence of the German Army was primarily a tactical air force until Anglo-American bomber forces compelled them to produce a strong fighter defence organisation.

In any event the speech made by Nikita Khrushchev in January 1960 before the Supreme Soviet portends a number of major organisational changes in the direction of Soviet armed forces because of the growing importance of submarines which can launch long-range atomic rockets, and of the declining importance of manned fighters and bombers particularly in the PVO Air Defence Command and the DA long-range attack command. Some of the possible changes have been hinted at earlier in this chapter. Unless the Soviet Air Force is given operational control of a considerable force of long-range rockets its influence in Soviet military strategic and planning councils is likely to decline during the 1960s. Perhaps its main function will be to be a chief administration at the Defence Ministry responsible for the inevitable reduction of bomber and fighter regiments and for an overall reduction in the scope of Soviet military aircraft manufacture, recruitment and training. Nevertheless because of the inevitable expansion of military transport air regiments and helicopter units, the continual need of air reconnaissance and, in my view, the need for conventional tactical aircraft if only to supply the allies of the USSR and to fight local secondary wars, the General Staff of the Soviet Air Force will continue to function in the 1960s on much the same lines as in the 1950s from an administrative point of view. The operations section will continue to be responsible for the basic training of air crews and for the operational training of the tactical air forces. It will however probably be more closely linked and more generally subordinate to the overall requirements of the Operations Division of the Soviet General Staff. The Soviet Air Force Intelligence Division has always been closely tied to the

Intelligence Division of the Soviet General Staff, but with the growth of rocket weapons and the consequent need for greater inter-service thinking, the work of the Soviet Air Force Intelligence Division is likely to become even less autonomous within the air force and less important. The same is likely to be true of the meteorological section in terms of autonomy, but its scope may well increase in view of the new weather knowledge being acquired in the course of Soviet excursions into outer space. The Air Transport section of the Soviet Air Force is likely to expand, but in view of the overall strategic importance of air lift in the future, it may well become a special administration directly subordinate to the Defence Ministry.

The main elements of the Soviet Air Force will continue to have separate administrations for such things as radio and radar, maintenance, air gunnery, instruments, navigation, airfield construction, medical matters, communications, strategy and tactics, personnel and publications; but the size of these departments is likely to shrink and fusion of some of them (e.g. navigation, communications and radio and radar) may occur within the next few years. Two aspects of staff administration which are unlikely to change, however, are the "special section" manned by the secret counter-intelligence officers and the chief political administration which is jointly subordinate to the Commander-in-Chief of the Soviet Air Force and to the Chief Political Administration of the Soviet Armed Forces. On the other hand the training of senior air force officers earmarked for command or staff careers is likely to undergo a change with a higher proportion attending combined service courses, such as are held at the Frunze Academy with more Air Force officers specialising in the rocket courses held at the Army's artillery academies.

Thus the 1960s will probably produce a modified re-organisation of the staff and administration of Soviet Air Forces. While some of the features of the 1950s such as the growth of the PVO air defence organisation and the separate administration for airborne troops are likely to remain, the growth of rocket weapons in the last five years will entail changes in the next decade, in which Army, Navy and Air Force will all be concerned. It would for instance not be surprising if a special arm of the services controlling all atomic weapons were established in the Defence Ministry.

If the Soviet military past is any guide to the future, then it is the Soviet Army and Navy which will acquire greater prominence in the

Soviet Defence Ministry. The Soviet Air Force may look back nostalgically to the 1950s as the period when the rise and development of Soviet manned air power gave its staff the temporary illusory promise of wielding greater authority in making the long-term military decisions which decide the strategy of the Soviet armed forces.

CHAPTER 16

INFLUENCE OF POLITICS ON SOVIET AIR POWER

THE relation between politics and air power needs no stressing. The career of the U.S. General Billy Mitchell, the story of Mussolini's adverse influence on the development of Italian air power in the 1930s are but two of many examples. It was clearly the strong political position held by Hermann Goering in Nazi Germany which enabled the Luftwaffe to develop more independently of Hitler and his party chiefs; but when the Luftwaffe failed in 1943-44, Hitler's interference and influence grew until in the summer of 1944 he temporarily diverted part of the Me 262 twin-jet fighter strength from its vital rôle in the air defence of Germany to the futile rôle of bolstering the tactical air support forces of the Wehrmacht.

It would be surprising if the Communist Party did not have more influence on the development of Soviet air power than politicians have had on the air forces of other countries. In the USSR the Party is the absolute ruler of the country and of its armed forces. It is no coincidence that Nikita Khrushchev has held the rank of lieutenant-general in the Soviet armed forces and that Stalin was the Generalissimo in command of the Red Army, Navy and Air Force in the Second World War. More than that, his military platitudes and inexactitudes have been ventilated regularly in Soviet military circles from the early 1930s up to his death in 1953. But although the links between the Communist Party and the Soviet armed forces have been so close and the Party has dominated some aspects of air and rocket policy, in an age in which the scientist and technician must play the major rôle in long-term weapon decisions, the real planning and strategy must be left more and more to them and to the military even in the USSR. The Party must concern itself more with such things as security, appointments and dismissals and political propaganda among the armed forces.

There are three main aspects of the interplay of Soviet politics and the Soviet Air Force. First the extent to which Soviet political doctrine and practice and the Party machine affect the daily life of airmen; secondly its influence on the design and production of aircraft, missiles and air and rocket equipment or on Soviet air tactics and doctrine; and finally the political supervision of the Air Force by the Communist Party aided by the political-cum-military police forces of the USSR.

It is the latter aspect which is by far the most important, absorbing many hundreds of thousands of men of all ranks in the work of counter-intelligence, record-keeping and supervision generally of Soviet Air Force conduct and promotion. The Communist Party is represented at all levels of the Soviet Air Force by a powerful network of administrators, officials and helpers. The Central Political Administration of the Soviet Air Force is housed in an impressive but old-fashioned six-storey building in a street just off the Gogolevsky Boulevard in Moscow. It is conveniently situated close to the buildings of the Soviet Defence Ministry and the Soviet Air Force directing administration. The officers in the Soviet Air Force Central Political Administration do not in general undergo any flying training or technical air or rocket courses, but their uniform is similar to the Soviet Air Force officers' uniforms, and in some cases they occasionally wear the full flying insignia and epaulettes.

The Central Administration receives its directives from the Central Political administration of the Armed Forces of which the head is called the Party *Apparatchik*, that is he is responsible for the rulings and interpretations of doctrine and current events. From this central organisation come the hundreds of instructions, orders and explanations in booklet, article or lecture-note form giving the party line on any topic of the day such as the dismissal of Defence Minister Marshal Zhukov, the launching of U.S. or Soviet earth satellites or Molotov's "anti-party orientation".

The Soviet Air Armies and Air Corps have minor editions of the Central Political Administration of the Armed Forces in Moscow. They run their own political newspapers or newsheets which are printed locally and they have a corps of Party agitators and propagandists, libraries of political literature, in fact the entire machinery of Marxist persuasion which is the very heart of Soviet civil and military administration. The machinery works something like this: the Central Political Administration of the armed forces circulates a directive to the air armies which is called an "orientation". It might run on lines such as this: "The Anti-Party thinking of Molotov (or Zhukov or indeed any Soviet leader destined to fall) has slowed up the re-equipment of the Soviet Armed Forces with new aircraft and rockets. This has retarded our progress in the defence of the Soviet peoples." This goes to the political administration of the air army, who work on the text, embellish it perhaps with one or two anti-Western sentences and send it to the political section of the Air Corps. They too revise it and elaborate the text before sending it to the senior political officer in the air division. Then it goes further to political deputies and to the commanding officers of air regiments. They discuss the much rewritten original "orientation" with Komsomol and Communist Party officials in the air regiments and at this stage it reaches the flying and ground personnel of the Soviet air Force.

The technique of political indoctrination varies. Sometimes the entire air regiment is assembled and the political deputy for the Communist Party makes a harangue lasting an hour or more on the chosen theme. Higher air echelons at the division or corps level might also send down a political officer to the air regiment who would discourse on the same themes more or less in the same way. In the course of the next week there might be in addition three or four so-called "political lessons" on the same subjects which might also be introduced into the information talks which are part of the Soviet airman's daily fare. In all these manifestations of political indoctrination it is considered important to retain the full and original meaning of the Party directive or message received from the Central Political Administration of the Armed Forces.

Clearly the Soviet believe that repetition is the soul of indoctrination and in this they do not differ from modern commercial advertisers in the West. Equally the political officers in the Soviet Air Force find themselves compelled to sustain themes which are manifestly absurd, such as, "The contribution of Comrade Stalin to the development of the Russian language". Nor indeed was the theme, "The contribution of Comrade Stalin to the development of Russian Russian military thought" very much less absurd. In public statements and in Soviet military books published before 1953 there is evidence to persuade one that Stalin's influence was paramount in the Soviet conceptions of doctrine and strategy. In 1948 for instance

Colonel-General Sudetz, then Chief of Staff of the Soviet Air Force and later Commander-in-Chief of the Soviet long-range bomber force said in a public speech: "In the Stalinist military theory it is considered that victory in contemporary war is attained only by the combined efforts of all types of forces. The development of the Soviet Air Force is carried out in accordance with this principle." Such was the lip-service paid to Stalin, though in fact there is nothing to support the view that Stalin ever influenced aircraft development or air operations. One prefers the more realistic views expressed by Colonel Tokaev. (*Soviet Imperialism*, Duckworth, page 30) who expressed the view that Stalin had an indifferent knowledge of tactics and strategy and accepted the rules drawn up for him by his experts. Tokaev was an active senior air staff officer in 1948 and he had direct knowledge of how Soviet military theory and practice were formulated.

After the formal talks and lectures by political officials might come the discussion and debate amongst the rank and file of the Air Force. Inevitably unrelated topics creep in to the discussion as Soviet officers, N.C.O.s and men wish to impress the Party and so facilitiate their promotion by adding their quota to the debate. The American "monopolists, cartel formers and colonialists", "British warmongers and slanderers", the reporting of minor delinquencies by other airmen (e.g. absence from duty because of a date with the girl friend, minor cases of pilfering such as are found in all armed forces), these are the kind of additional irrelevent topics thrown in to demonstrate the individual airmen's loyalty to the party and its doctrines. These are a necessary *hors d'œuvre* in the pattern of daily life in the Soviet Air Force. They are part and parcel of the training and study system in force usually in the evenings when the daily flying or technical training is over. Evening political studies take various forms, for instance the kind of lectures, talks and discussions already mentioned, political study groups and Communist Party or Komsomol meetings.

It is extremely difficult to assess the impact of these party political activities on the ordinary run of Soviet Air Force officers and men. Most airmen in all air forces have a healthy disregard for doctrine and propaganda and do not betray a very profound interest in current affairs as seen by the Communist Party or indeed from any other standpoint. The writer has had two types of personal experience on the subject. The first was during the Second World War from August 1941 to November 1944, during which period some fifty Soviet Air Force officers ranging in rank from Captain to Brigadier-General whom I met regularly, had the opportunity of showing the Party indoctrination in the course of a wide range of discussions on war in all its aspects—sport, family life, eating, love-making and current affairs. But no attempts at propaganda were made on behalf of Stalin, Marx or the Communist Party even by those Soviet Officers who were professionally part of the political administration of the Soviet Air Force. There was plenty of patriotism and military discipline, a favourite opening remark was "Our general staff says that . . ." But in the hundreds of discussions on strategy, tactics and war planning I cannot recall a single reference to Stalin or Lenin or Marxism. And when I tried to discuss them with the Soviet officers, they smiled and said: "These things are not important for our common victory against the enemy!"

One got a similar impression of the unimportance of the political indoctrination in reading the diary of Flight Lieutenant Hubert Griffiths, in civilian life an author and critic, who was in the last war adjutant to the Hurricane fighter wing which lived and fought with elements of the Soviet Naval Air Force on the Murmansk front from September 1941 to December 1941. Griffiths was Russian-speaking, as were other R.A.F. officers in the Hurricane Wing. They lived closely with this group of Soviet naval airmen. They flew with them in combat, ate with them and discussed aircraft technicalities, the weather, tactics and problems of administration. But the author of the book in his assessment of Soviet flying capabilities, comradeship and personalities, does not even mention the political element. No doubt the R.A.F. wing were briefed to avoid the subject, but Griffiths' critical observance seems to have detected no party interference in the process of training the Soviet pilots to fly British Hurricanes or with the Soviet ground technicians who had to assemble, maintain and service the machines.

But the most convincing and important evidence of the relative unimportance of Party influence in the day-to-day military affairs of the Soviet Air Force emerges in the various accounts of the Normandie-Niemen combat operations and co-operation with the Soviet Air Force in the Second World War. There are two long books of intimate memoirs on the subject and in addition the author has had a number of direct talks with General Pierre Pouyade who commanded de Gaulle's fighter squadrons on the Eastern Front. General Pouyade had to deal with the Soviet air authorities mainly at air regiment and air division levels, and while aware of the disciplinary efforts and the counter intelligence organisation of the Soviet Communist Party system, he did not feel that fundamental air matters, such as the supply of aircraft, tactics, decisions on operations, combat readiness, etc., were subject to more than the cursory occasional influence of local political officers or party officials. And reading through the accounts of air fighting on the Eastern Front published by some of General Pouyade's pilots (e.g. Roger Sauvage's *Un du Normandie*) one finds that the political police get only passing mention and that the general day-to-day life of Soviet airmen in war scarcely seems to be affected by the impress of the party machine. A similar impression emerges from the German studies of air operations on the Eastern Front in World War II.

However at the higher level of the direction of air operations, i.e. at air army level and in the confines of the Kremlin, the interference of the Communist Party machine was more frequent. It is of interest that Luftwaffe General Walter Schwabedissen in his appreciation of Soviet Air Force staff work in the summer of 1941 described it as "lacking flexibility and occasionally hindered by Communist Party control". (See *Soviet Air and Rocket Forces*, Weidenfeld and Nicolson, page 48). Remembering the great pressure on the Soviet Air Force during this period of the Germano-Soviet War perhaps the most interesting aspect of the General Schwabedissen's appreciation was the use of the word occasional. Nevertheless the association between the commanders of the air armies and the senior political air officer was intimate and continuous. Major-General Ramazanov for instance was the political member of the military council of the 2nd Air Army in World War II. Although he cut an unimpressive figure besides its Commander-in-Chief, Colonel-General of Aviation Krassovsky, nevertheless whenever Krassovsky went to an important meeting or took an important decision, he not only took Ramazanov with him but was careful to get his advice. This was typical of the close relationship which existed between air commanding officers and political party officers not only at army but also at corps, divisional and regimental levels.

And after the war, Stalin and his Politburo certainly interfered actively with Soviet long-range bomber policy and also with long-range rocket policy. When Stalin personally saw the effect of strategic bombing in Berlin during the Treaty of Potsdam negotiations in 1945 and learned of the German blue print for a transatlantic rocket and rocket bomber he ordered the revival of the then moribund Soviet ADD Long Range Bomber Command, gave it a new lease of life and in 1946 re-established it with a new title, the DA (initials mean "long-distance aviation"). It was given a high production priority, so that by Stalin's death in 1953 it had grown into a force of about a thousand four-engined bombers. It was Stalin too who put emphasis on the early development of Soviet long-range rockets and laid part of the foundation of the technical lead in this field which the USSR presently enjoys over the West.

But was he exercising any more control over defence policy in the USSR than is exercised in the United States by the President and his military advisers, or in England by the British Cabinet and Chiefs of Staff? The U.S. decision to continue to build aircraft-carriers or the British decision to create a V-bomber force must in the final analysis be made by politicians. Once that decision is made the military and the aircraft or rocket industry are left to work out the research, development and production details. The same sort of thing happens in the USSR as in the United States or Great Britain. There is nothing to show that the Soviet aircraft designer Tupolev has suffered political interference in his post-war development of Soviet jet bombers and transport planes, and the same applies to other aircraft designers such as Antonov, Mikoyan and Gurevich. Nor is there any evidence that the designers of long-range rockets such as Yakovlev and Pokrovsky have been influenced by the Communist Party machine unduly.

It is virtually impossible to unravel the conflicting political elements of the evidence of the influence of Soviet politics on the strategy and high direction of war in the USSR. But then this is equally true in Western democratic countries as all well know who have read the inconsistent picture which emerges from the war memoirs of President Eisenhower, Lord Montgomery, Sir Winston Churchill and Field-Marshal Lord Alanbrooke. In trying to isolate the elements of truth the mind has the impression of walking along a sagging tight-rope. The Soviet General Krasilnikov wrote a few years ago: "Military strategy is directly dependent on politics to which it is subordinate. Strategic war plans are worked out on the basis of objectives established by politics." Clearly there is truth in the last sentence, and the subject of the Hungarian Revolt in 1956 is a classic example of politics providing the objective.

But Mr. Khrushchev has in the last few years blown much fresh air into the corridors of Soviet military and political thinking. At the Twentieth Party Congress of the Communist Party in February 1956 he said that the Marxist-Leninist precept on the inevitability of war had been correct for its time but no longer corresponded to contemporary conditions. Certainly in Khrushchev's reign at the Kremlin there has been abundant evidence that there has been a lessening of political control over military affairs in the USSR and a more realistic approach to problems of war without any important lessening of the control of the Communist Party over the armed forces. For example a number of political officer posts in the lower echelons of the Soviet Army and Air Force have been eliminated and the general importance of the political administration in the Soviet air and rocket forces has declined since 1953. Some of its previous functions have been taken over by the military and air force personnel and others by the Party and Komsomol organisations. Marshal Zhukov, the ex-Soviet Defence Minister, is credited with having achieved this reduction in the rôle of Soviet political officers. He was probably mainly responsible for the April 1957 directive to the Soviet Army and Navy (which also applied to the Air Force) that the "commander guides the entire military and political training of the personnel".

However since Zhukov was ousted from his position as Defence Minister the political officers have tended to increase their influence in some of the military councils which discuss regional military policy in the military districts of the USSR. The influence of the KGB, the Committee for State Security which is separate from the armed forces, is still paramount and separate from the Ministry of Defence, and its counter-intelligence operations keep a close watch on the activities of the personnel of the Soviet air and rocket forces.

It would seem that the 1950s was a period of fluctuating political influences in Soviet army, navy and air affairs. It divides into the Stalinist period up to 1953 and the Khrushchev period since. There is no doubt that Khrushchev has allowed Soviet military leadership to acquire much more control over high-level planning and military strategy than did his Georgian predecessor. Stalin had tried to establish a dictatorial ascendency over Soviet military strategy and thought. But I do not think he succeeded, except to limit the amount of independent published doctrine and viewpoint from the pen and voices of his air, naval and military commanders. It is true that in general Soviet published air doctrine from 1945-53 tended to adhere to the Soviet pattern of World War II which limited air forces to the rôle of supporting the ground forces. But all this time there was a re-evaluation of air weapons going on in Soviet Air Force schools and staff colleges, in the aircraft and rocket factories and in the informal discussions in Soviet Air Force circles. If the new doctrines were curbed by Stalinism, they peeped out here and there in the Soviet military press. Thus writing in the Soviet Air Force magazine *Herald of the Air Fleet* in 1946 Major-General of Aviation Tatarchenko surveyed developments in atomic weapons, radar, jet-propulsion and rocket missiles, and made the point that a greater rôle would be given in the future to strategic bombing. He wrote cautiously: "It would appear that contemporary air forces are capable of carrying out not only tactical but also strategic tasks which no arm other than aviation can fulfil."

Another way in which Stalin kept control over free thinking by his air leaders was to purge them or prematurely retire them. He did not in the immediate post-war period revert to the large-scale purgings of the 1930s but he saw to it that many senior air force officers with a proud record of commanding air armies and conducting operations against the Luftwaffe and the Wehrmacht were removed from the Soviet Air Force High Command. The best known was Chief Marshal of Aviation Novikov and his wartime deputies, Marshals of Aviation Vorozheikin, Khudiakov and Astakhov. Other prominent Soviet air leaders who were retired unexpectedly were General Repin, head of the Engineering Service of the Air Force, and General Grendal the head of Air Force Intelligence.

The rise of Marshal Zhukov as Defence Minister under Nikita Khrushchev marked a new phase. He achieved some reduction in the rôle of the political officers in Soviet army, navy and air affairs and also managed to secure greater dominance of the military hierarchy by making Soviet strategic planning essentially a technical and professional question and limiting the interference of the Party to some extent. But his dismissal in 1957 suggests that he was trying to go too far. The reasons for Zhukov's dismissal are not completely known but he was significantly not on the list of those who were to be purged by Stalin early in 1953. There is reason to believe that during this so-called period of the "doctors' plot" Khrushchev was a supporter of Marshal Konev rather than Marshal Zhukov. Perhaps Khrushchev was simply biding his time before dismissing Zhukov, who as one of the best military brains and most popular figures in the USSR had to be treated with some caution.

As with so many other Soviet internal political crises one cannot unravel all the strands of the doctors' plot, first announced in January 1953, in which prominent Soviet physicians were accused of having actually murdered Zhdanov and Shcherbakov, and of plotting to kill a number of Soviet military leaders. How far military and air force leaders were involved in the plot or were likely to be victims of a coming Stalinist purge is hard to say. Stalin's sudden death at the time will no doubt prevent the complete story being told. Certainly when Stalin died in March 1953 there was inevitably a shake-up in Soviet military and political leadership and in this period Soviet military leaders seemed to acquire rather more authority in Kremlin councils. Marshal Zhukov had been named a Deputy Minister as Commander-in-Chief of the Armed Forces at the end of 1952 and after personally arresting Beria, head of the Soviet Secret Police, in the spring of 1953 he became a first deputy Minister of Defence under Bulganin's Ministry. Then in February 1955 Marshal Zhukov replaced Bulganin as Minister of Defence and was himself replaced by Marshal Malinovsky who still survives.

Marshal Zhukov had obtained more authority within the Soviet armed forces than any other Soviet military leader since Marshal Tukhachevsky in the 1930s. One month after he had become Defence Minister, twelve new marshals and generals were created, most of them former military associates and friends of his. By abolishing the political officer at company and squadron levels he was reinforcing the authority of the professional army, navy and air commander, and he continued to press for further authority against the pressure of the Party. Two points must be made however. Although in the detailed hierarchy of the army, navy and air force there was some debate in Zhukov's day about the precedence of the Party or the military there can be no question at all that the Communist Party will always play and has always played the dominant rôle in national military policy—that is the question of peace or war. It could be that one reason for Marshal Zhukov's dismissal was that he wanted to play too great a rôle in the Soviet military decisions at the time of the Hungarian Revolution in 1956. The second point is that the Soviet Air Force High Command seems to have played no very important part in the political manœuvrings of the 1952-57 period, thereby confirming its secondary status in Kremlin defence policy making.

But the biggest political influence internationally on the Soviet Air Force in the last five years or so has come from the development of the new rocket weapons and not from any changes of political or military leadership, and this influence will be even more pertinent in the coming 1960s. The amassing in East and West of inter-continental missiles and of strategic and tactical atomic weapons has produced an atmosphere of greater stability for disarmament discussion, which will in my view be successful and gradually cause a reduction in Soviet air and rocket forces. The Soviet authorities will no doubt retain the KGB organisation to watch over Soviet Air Force and other military personnel and the Party machine will infiltrate with its pep talks and directives and pamphleteering activities at all levels of the Soviet air and rocket units. But when one looks back at Soviet military policy in the 1950s one will see that it was in the first place almost entirely influenced by the inevitable weapon developments achieved by Soviet scientists and engineers and not by any hothouse school of Stalinist or Marxist doctrine or any political pressures from Mr. Khrushchev. The Soviet leaders' speech on Defence Policy early in 1960 might have been largely the result of a White Paper conceived in London. It put the emphasis on both tactical and strategic atomic weapons, on missile-launching submarines and on a reduction of conventional ground and air forces. In fact the entire Soviet development in air and rocket power since 1945 has had a military logic and sequence. Whatever Stalin and other politicians have said or thought about strategic bombing, atomic weapons, helicopters and rockets, there have been logical and inevitable weapon developments in the USSR, on which Soviet politics has cast its shadow from time to time but without exercising the direct influence of a designer like Tupolev or a rocket scientist like Pokrovsky. We live in an age in which neither the politician nor the military leader is the seer of strategy and doctrine. The politician can remove military and air leaders and attempt to infuse standards of security, political indoctrination and discipline. But it is the weapons themselves, the standards of technical education of personnel and the quality of technical production of equipment and checking and servicing rocket and missile parts which are assuming greater proportions. A cynic might well ask if there will be room for KGB or party officials on the first Soviet military earth satellite.

But the really important thing is that the Party has a close control over the distribution and use of all atomic weapons. On occasions of international political tension, when brinkmanship is being exercised perhaps on both sides and a minor military incident could flare up into a world disaster, the NATO authorities may be glad that the restraining influence of the Communist Party over local military Soviet leadership is strong. What so many people are afraid of is that a Third World War could happen as the result of a local mistake. The dominance of the party machine in Soviet military affairs lessens the likelihood of that mistake. If Soviet Marxism has tended to discourage individual initiative it may also be a key check on individual foolishness in the thermo-nuclear weapons age.

Although in fact science and technology must be the main influences on Soviet military development now and in the future, one cannot hope that even the best of Soviet military scientists and engineers can divorce themselves from lip-service to Marxist statements and to anti-capitalist pronouncements. General Pokrovsky illustrates this inevitable tendency to drag anti-Western pronouncements into a military context even when it requires a conjuror's effort to do so. In his recent book, *Science and Technology in Contemporary War*, he deals logically and objectively with weapon developments but has to introduce a political *hors d'œuvre* of utter irrelevance such as "Dialectical materialism arms scientists with the correct philosophical view of the world" or "The superiority of Soviet military science over bourgeois military science was especially evident in the Great Fatherland War of 1941-45." Despite these interventions of blatant propaganda the book discusses the range of modern weapons from radar and rockets to vertical take-off planes and rocket-firing submarines with all the critical analysis and reasoned doubts that might be expressed by a Western military writer.

The fact is that underneath the superficial trappings of Communist propaganda and Soviet Marxism, the re-thinking of Soviet defence policy is as urgent and as hesitant as that going on in NATO and London and Washington. Politicians and budgets will in the future influence decisions and strategy in every country in the world. The Soviet political police and the easier machinery for removing military leaders are the main differences between East and West armed forces as far as politics are concerned. But on the more important question of peace and war, it will gradually be found, in my view, that the U.S.A. and the USSR have much more common ground than they think they have in 1960. We live in an age in which politics have debased the currency of words and ideas, as Aldous Huxley predicted in *Ends and Means* more than twenty years ago. One of the main political tasks of the future is to restore the currency from the standards of meaningless speeches to those of worthier metal.

CHAPTER 17

FUTURE PROSPECTS

A LARGE-SCALE war between two world powers is no longer "the continuation of national or international politics by other means" for it could create chaotic conditions in haphazard fashion, conditions over which neither side could exercise political or military control. Any minor incident between the forces of NATO and the Warsaw Treaty powers is capable of escalation to the maximum use of atomic weapons on both sides and without either retaining control over the developing situation. It would call for rapid decisions based on uncertain or conflicting information. That is why one believes that the Soviet proposals for eventual disarmament are genuine. And yet even they must know how impossible it is to achieve a perfect inspection system which may call for the examination of every nook and cranny and the pulling up of almost every floor-board. There is also the growing urgency of bringing Communist China and other powers such as France, who may acquire atomic weapons in the 1960s, into any international agreement for comprehensive disarmament. But until there is clear evidence that disarmament on a world-wide scale is likely to be a practical proposition, we must expect that the USSR will continue its present military policy of modernising its armed forces and developing more and more powerful weapons.

Virtually the only authoritative statement on Soviet future defence policy in the 1960s was made by Nikita Khrushchev in a speech to the Supreme Soviet in January 1960. In this he said that Russian military planners worked on the assumption that a future war would involve the mass use of nuclear weapons, that he thought that one hundred or so nuclear weapons could reduce to rubble the main centres of an industrialised country roughly the size of England or France. He stressed the point that defence capacity is nowadays no longer a question of the number of men under arms but of fire power (surely this has always been so), and he added that despite the reduction of the Soviet armed forces since 1955 by one-third, and a further proposed cut of 1,200,000 men to a figure in a year or so's time of about two and a half million men under arms, USSR armed forces would in no sense be weakened by the reduction. Khrushchev added that the USSR had a sufficient quantity of atomic and hydrogen nuclear weapons to deal with any eventuality. He also dealt for the first time in a public statement with the question of a surprise attack which he was inclined to disregard as a likelihood. He thought that one could not assume that even if a surprise blow could be struck with nuclear weapons it would be able to prevent effective retaliatory attack. "It will always be possible to bring reserve means into operation and hit targets from reserve positions" he said.

Prior to the Khrushchev speech this was broadly speaking the view expressed by Dr. Dinerstein in his *War and the Soviet Union* (Atlantic Books, London) and there is no reason to believe that this is not a basic tenet of Soviet military and political thinking. It is no doubt conditioned by their superior ability to mask the location of many of their long-range rocket sites from Western intelligence sources. But perhaps a more important factor is the successful testing of a Soviet ballistic missile with a range of 7,500-8,000 miles which is apparently an accurate weapon capable of landing within a couple of miles of the target. Indeed there are some Western defence experts who believe that the USSR can extend the range of its intercontinental ballistic missiles to as much as 12,000 miles and that this will be the ultimate earth-launched Soviet weapon. Either of these two new Soviet ballistic missiles could be the incredible weapon referred to by Mr. Khrushchev in his defence speech of January 1960.

If successfully developed within the next two or three years, either of the two new long-range rocket weapons would give Soviet strategic attack planning a number of important new advantages. It would mean that the vast ballistic missile early warning system, with its long-range anti-missile radar chain, at present under construction in such places as Alaska, Greenland and Britain, would have to be radically extended at an enormous extra cost. This U.S. defence scheme is based largely on the postulate that the USSR would be compelled to fire most of her long-range missiles aiming for North American targets across the shorter Arctic routes. But if, as is believed, the range of Soviet rockets is being extended to 7,500-8,000 miles without any important loss of accuracy it means that the

USSR will be able to abandon the Arctic circle as a major area for launching her long-range rocket attacks. And since she is clearly going to phase many of her manned bombers out of her DA long-range bomber command in the next few years, all the difficulties of supply and maintenance of both long-range bomber and rocket units in the Arctic will be avoided. It would be surprising if there is not a major redeployment of Soviet long-range bomber and rocket units in the next year or so away from the Arctic bases to warmer areas nearer the Baltic and the Caspian Sea, and to the Eastern Maritime Provinces.

When Mr. Khrushchev said in his January 1960 speech that Soviet military aircraft are to be "almost entirely replaced by rockets" he was, in my view, referring to a gradual process which will take place in the 1960s and was thinking more of strategic manned bombers in the first instance. Nevertheless he did say explicitly that the air forces and the fleet have lost their former importance and that these forces would not be simply reduced but eventually replaced. He went on to say that the USSR might stop the manufacture of bombers and that the Soviet armed forces had been switched to rocket weapons which would be improved and developed further until they were banned under a general disarmament arrangement. There is no reason to doubt the validity of the broad trends of Mr. Khrushchev's 1960 statement on Soviet defence policy. Indeed most of the evidence of the five years that preceded it prepared the ground for his comprehensive review of that policy.

However one has to speculate considerably in offering even a broad prediction of what the new Soviet policy of the 1960s will mean in terms of the reorganisation and contraction of the Soviet Air Force. Estimates by NATO and other sources of Soviet aircraft operational strength have in the past few years ranged round a figure of 20,000 aircraft. This figure will almost certainly be reduced in the next year or so perhaps to some 15,000 aircraft, and by the middle or late 1960s to some 10,000 aircraft. It is however difficult to envisage a situation apart from general disarmament in which the Soviet Air Force would be reduced much below that strength. As long as the NATO or U.S. forces have a force of manned bombers, it would be logical for the Soviet PVO air defence command to retain a reasonable sized force of manned fighters to intercept enemy bombers from peripheral bases outside the USSR. The range and number of Soviet ground-to-air guided missiles will probably increase in the 1960s to a strength of many thousands and no doubt the PVO forces inside the USSR will consist almost entirely of these weapons. But there will be occasions when fighter aircraft are needed to investigate oncoming flights by aircraft which could be simply civil transport planes of which there will be many foreign nationalities operating over the USSR in the next few years. Moreover the current range of most Soviet ground-to-air defensive guided missiles is believed to be less than 100 miles and so the supersonic MIG and SUKHOI fighters still have advantages in radius of action and flexibility of use against bomber attack compared with the guided missiles.

There is also another basic reason for keeping a Soviet fighter defence force in being and that is to train the air allies of the USSR who are not likely to have mainly rocket defence forces in the 1960s. Politically it is hard to see whether the Soviet fighter aircraft commitment will expand or contract under this heading in the 1960s. It is by no means certain that China will be able to manufacture all the jet fighters she needs by 1970 and she will probably continue to rely on Soviet output for at least another decade. She may also feel the need to build up her metropolitan strategic fighter defence force. At present it is a negligible military asset and the call on Soviet fighter production in the 1960s for her allies may be greater than in the 1950s. The same may well be true of the needs of the United Arab Republic. And with the present trend of Soviet political cultural and economic penetration of Latin America, who can say how far those republics will draw on Soviet fighters in the next five or ten years? The recent announcement that Cuba has purchased MIL helicopters from the USSR may well be followed by news of purchase of MIG fighters. It would be surprising if the purchases of Soviet military aircraft by Latin American governments were confined to Cuba. And if in 1960 the emerging African territories show no sign of establishing military air forces, is it too fantastic to imagine that African leaders envisage the possible creation of an air force to put pressure on the South African government? Certainly the Kremlin must expect a demand for both fighter and bomber aircraft from a whole new range of potential military allies and so must keep in being the framework of a Soviet air force which can train pilots and aircrews from many other countries.

But the main argument for retaining fairly large manned fighter forces for the next few years has already been given in the chapter on tactical air power. It is believed that the Soviet defence authorities must see the importance of getting their airborne forces into action with the maximum initial degree of safety. Here it would seem that fighter and fighter-bomber aircraft have a long-term rôle to fulfil, in denying effective interception by enemy fighters and in strafing and smothering the full and effective employment of enemy tactical atomic weapons. If there were any evidence that the cut announced by Khrushchev in Soviet armed forces involved a major cut in Soviet airborne forces one would have to revise this view, but tactical fighters and airborne forces go hand in hand, especially in the age of major potential employment of tactical atomic weapons.

The Soviet success in perfecting long-range rocket weapons may well eliminate the entire DA long-range bomber command in the 1960s but once again one is inclined to think that at least a rump long-range bomber air force of a couple of hundred jet bombers will remain. Who can say in what circumstances the USSR might want to provide China with such a force and alternatively to fly armed reconnaissance at short notice over Chinese or Indian territory? Long-range rockets may be the finite strategic weapons of earth-based war, but they are not a means of urgent reconnaissance or investigation. Even more important is the uncertain prospect of future submarine or anti-submarine warfare. Let us assume for instance that some means of long-range detection of enemy submarines is invented. Is it not likely that on some occasions a supersonic bomber informed of its position could make an attack more quickly than a submarine? Furthermore as long as the U.S. Navy has a strong force of aircraft carriers capable of nuclear and thermonuclear strikes, surely the USSR is likely to retain a force of long-range fast counter-strike bombers which can attack them so much more quickly than the fastest submarines.

That is why one feels that on balance the USSR will go ahead with its present programme of developing both a long-range supersonic jet bomber and a nuclear-powered long-range bomber reconnaissance-cum-transport aircraft. The development of a force of supersonic long-range jet bombers in the Soviet Air Force has already been reported and is likely to continue for the next four or five years at least. The emergence of an atomic-powered aircraft some time in the 1960s is more than a likelihood. It is now more than six years since the first reports of atomic-powered aircraft engine developments were received. The Soviet project was stated to have had the help of Polish and German rocket and nuclear scientists. It must be apparent to all that the USSR no longer needs such help. The reported Soviet intention is to build an experimental bomber with a top speed of perhaps 1,000 m.p.h. and an endurance of more than twenty-four hours. The nuclear rocket motor is thought to have a thrust of about fifty tons. There have been occasional reports to suggest that the Soviet are making good progress with their nuclear-powered aircraft, including one unconfirmed report of successful flying trials a year or so ago. But there is no firm evidence that the USSR has solved the problem of producing an atomic-powered aircraft engine which can be shielded by sufficiently light material to protect the crew from its harmful neutrons and gamma rays. One report suggests that a relatively light material weighing about five tons is being tried out as a shield with reasonable prospects of success. While one can understand the political and technical prestige of producing the world's first successful nuclear-powered aircraft, it would have no great advantage in future military operations over a long-range supersonic jet bomber which could be flight-refuelled either as a bomber reconnaissance plane or as an airborne radar or radio counter-measures station. One is more inclined to think that Soviet nuclear-powered engines will be used for launching manned vehicles on interplanetary flights in the future rather than for powering manned aircraft. There appears to be one exception to this. Soviet military and naval writers in the past few years have mentioned atomic-powered aircraft as a kind of substitute for aircraft-carriers used in long-range ocean bomber reconnaissance patrols. *Military Thought*, an authoritative Soviet military journal, carried this specific remark towards the end of 1955: "The use of atomic propulsion for naval aviation makes it possible to attack the enemy in support of the fleet from land bases without the use of aircraft-carriers." And as the number of Soviet surface ships declines in the 1960s and the number of ocean-going submarines increases, so will the potential value of long-range naval patrol bombers. Thus in addition to the political and prestige value of an atomic-powered aircraft it may become an essential operational requirement of the middle 1960s for naval air operations. It is of some interest that the Soviet have so far spoken mainly of land-based aircraft with nuclear propulsion.

It would however not be surprising to learn that they have been experimenting with Beriev flying-boats in the field, just as the U.S. Navy have with the Martin Seamaster flying-boat.

It is reasonably likely that one or perhaps two of the three Soviet long-range bomber armies will be disbanded within the next two or three years and that the output of long-range jet bombers will be substantially reduced. On the other hand the expansion of the military and civil air transport organisation and resources of the USSR is likely to be the most outstanding aeronautical development of the 1960s. One wonders if a combined military and civil air transport command will not in fact be established. Some of the aircraft production capacity, manpower and tools which were allocated to the manufacture of jet bombers in the middle and late 1950s will be turned over to the production of jet transport planes in the 1960s. In both the Aeroflot air lines and the military air transport regiments, the non-jet machines will be replaced by turbo-prop and turbo-jet twin- and four-jet machines such as the Tu 104, the Antonov 24 and 10, and Ilyushin 18 and the Tupolev 114. Since the Tu 104 is in part a derivative of the Badger jet bomber and the Tu 114 giant four turbo-prop passenger plane a descendant of the Bear bomber, the next year or so is likely to see a further acceleration in the output of long-range jet transport machines. The process of dovetailing the decline of the Soviet long-range bomber force with the expansion of the long-range jet transport arm will now become much easier.

It seems likely too that the USSR will develop a transport version of the supersonic delta-wing bomber which was reported to be flying in 1958 and entering long-range bomber service in 1959. There is every incentive for Aeroflot, to be the first civil air line to operate supersonic passenger aircraft. Since 1955 Soviet civil aviation has energetically pushed forward its negotiations to operate its new jet planes outside Communist territories. At first there were agreements which enabled Aeroflot air liners to fly to Vienna, Stockholm and Copenhagen. Then came extensions to Paris and London. There can be little doubt that this overseas route expansion will reach out in the 1960s to the airports of Asia, Africa and Latin America, and probably North America. For the first time in aviation history Soviet pride need not suffer when her aircraft are parked side by side with the best Western passenger planes on the airfields of foreign countries.

On the internal air lines of the USSR one of the big developments of the 1960s will be the expansion of the city-to-city jet helicopter services already started. Up to 1960 the bulk of large Soviet helicopters was used for military purposes either airborne troop- and equipment-carrying operations or on military communication flights or on agricultural work with Aeroflot. But with the large-scale manufacture of the twin-jet MIL 6 helicopter which will carry more than 70 passengers and can carry up to 100 troops, rapid development of passenger helicopter services, particularly in the Crimea, is likely. The new Soviet jet helicopters, like the long-range jet passenger air lines will almost certainly provide commercial competition in Asian and Latin American markets for British, American and French aircraft and helicopters in the 1960s. Once the USSR establishes confidence that they can provide a satisfactory after-sales service and are willing to make the necessary modifications to suit the customer they could become major competitors to the West in the international sales of air liners, helicopters and airfield equipment. They will probably make their initial attempts in such countries as Indonesia, the United Arab Republic and perhaps Cuba and Brazil with an offer to sell at prices much below those of Western competitors. In addition there may be a Soviet offer to help in the construction or modification of airports, to install navigational and landing equipment and to maintain and service the aircraft sold.

Side by side with the Soviet expansion in civil aviation will be a continuing expansion in the military transport arm which may well give the USSR the largest air lift resources in military aviation history. If, as Khrushchev has said, the armed forces of the USSR are to be substantially reduced without any loss of firepower or military effectiveness then it is likely that the airborne forces will be expanded and with them the military air transport units. It would be reasonable for the USSR to double its élite parachutist airborne infantry forces in the 1960s and to have a force of at least 2,000 long-range jet transports not only to lift perhaps five divisions in one operation but also to transport by air large numbers of tactical atomic weapons which in peacetime would probably be mainly kept back in metropolitan areas of the USSR. Like the Western powers the USSR must be ready to have fire-brigade airborne forces and they cannot know how many places they may have to be flown to in any given period. There could be need for them on the borders of China and India, in Iraq, eastern Germany, on the Persian border. Moreover since their road communications in war could be so easily disrupted by atomic attack the despatch of airborne troops may be the only practical way of bolstering up a local ground situation. We may well be in an era in which road and rail communications are no longer a prime long-term factor in military planning of movement in and behind battle areas. Moreover the great advantage in developing air transport resources to the maximum is that if disarmament comes and the USSR has to beat her swords into ploughshares, the military transport planes and helicopters will be most readily convertible to important economic use in an era of peace which would intensify political and trade competition between East and West. Trade penetration of Africa and Asia and indeed cultural and political penetration would be speeded up and facilitated by supersonic and other long-range air communications. Large-scale modern air transport may also be a vital element in any efficient disarmament inspection scheme on an international scale. If when the scheme is in force Soviet air transport resources, say in the late 1960s, have a preponderance over the rest of the world, they could achieve thereby a major political advantage.

Meanwhile reorientation of Soviet military air power must be a prior preoccupation in the Kremlin. Among the overall changes of the 1960s a major overhaul of naval air power is probable. As in the case of the PVO strategic defence forces, the fighter regiments defending ports are likely to be replaced by ground-to-air guided missiles and some of the twin-engined Ilyushin 28 and Tupolev 14 medium-bomber jet squadrons are likely to be replaced by a force of long-range four-engined jet bombers, at first subsonic then supersonic and perhaps later atomic-powered machines. This could mean that the operational strength of Soviet naval air forces will probably be reduced by at least 1,000 aircraft in the next few years. A new conception of Soviet strategic naval air power may develop in the 1960s in which long-range jet bombers designed by Ilyushin and Myasischev and Tupolev will be responsible for hunting and attacking Western submarines and aircraft carriers and taking part in long-range attacks on enemy-held airfields and rocket bases. Thus the Russian naval air squadrons which were the first to be equipped with long-range bombers in the First World War may become the last Russian air squadrons to be re-equipped for a strategic air rôle. The new long-range bombers would also be expected to carry out mine-laying and attacks on enemy shipping in any war in which strategic atomic and thermo-nuclear weapons were not in use.

But it is Soviet submarines which will by the late 1960s be making a major contribution to the Soviet strategic air potential. It has been noted by Admiral Burke that over half the major cities of the United States lie within 100 miles of the coasts. Clearly with the development of Soviet nuclear powered submarines and of submarine-launched IRBM Komet and Golem nuclear rockets, the Navy will assume a greater rôle in the Soviet overall strategic deterrent force. Soviet submarines are known to be active both in the Pacific and the Atlantic mapping the suboceanic terrain. There are even reports that they are planning permanently manned undersea missile stations to be supplied by large transport submarines and also undersea missile-containers which can be anchored by submarines and later fired by remote control. Such weapons would be almost impossible to detect.

There are some military thinkers who believe that for the next few years the bulk of Soviet deterrent forces will consist of the growing numbers of intermediate-range and intercontinental ballistic missiles which will be land-based and widely dispersed in the USSR so that detection will be extremely difficult. Thereafter in the late 1960s and 1970s the development of military sputnik earth satellites will constitute the major new facet of Soviet military strategy. These would be used in all three major rôles of reconnaissance, attack and defence. By using infra-red, telephoto and other devices on earth satellites the USSR may for the first time achieve a completely accurate map of the globe and may be able to keep watch on important military movements in all parts of the world. There are those however who believe that as an earth satellite has to orbit several hundred miles above the earth the photographic equipment will never be able to detect such things as rocket sites. Perhaps with more powerful rockets and other devices such as retro rockets the USSR may be able to produce earth satellites that can orbit at lower heights.

Earth satellites can scarcely be used as a major element in Soviet strategic air defence though they may be used to jam the initial radar guidance of enemy rockets. But the Soviet PVO of the 1960s will go through two evolutionary processes. First the change from manned fighters to anti-aircraft guided missiles, and then the more important change to anti-rocket missiles. While the PVO strategic air defence system may well have the use of long-range anti-missile radar to detect the path of oncoming rockets, it is difficult to see how they can develop an effective anti-missile interception system in the foreseeable future. There are of course many potential weapons and devices which can be tried out. In theory they might use electronic techniques for changing the path of enemy rockets in flight, especially if they could mount such radio counter-measures from earth satellites. Soviet scientists might achieve a breakthrough in being able to use the force of gravity to turn aside oncoming rockets, though at the moment they are at the stage of observing the effects of gravity rather than harnessing it to military defence. The PVO might well develop a new type of anti-missile barrage which could create concentrations of artificial meteorites which would divert hostile ballistic missiles before they re-entered the earth's atmosphere. All these things are possible, but at the moment there is no reason to believe that the USSR has any means of effective defence against long-range missile attack. It is no doubt this military inability which makes many experts think that there is a Soviet plan to launch a surprise or preemptive long-range missile thermo-nuclear attack on the U.S.A. and Britain.

Perhaps the USSR is also planning a new deterrent long-range rocket to be fired from earth satellites. Following the intercontinental ballistic missile we may have the interplanetary ballistic missile. Sputnik space platforms could in theory and eventually in practice be armed with ballistic missiles which can be fired at specific target areas on earth. They could be unmanned and fired by electronic control from earth and eventually manned and probably given in this way greater accuracy. But the interplanetary ballistic missile is not likely to be a weapon of the 1960s, though technically unmanned but rocket-armed space platforms or simply armed rockets orbiting in space are a possible development of the 1970s. But almost certainly the development of submarine-launched missiles, of more accurate and longer-range land-based missiles and of anti-missile missiles will have priority over military space platforms in the next decade.

Of these three weapons, the greatest challenge to Soviet ingenuity will come from the requirements of anti-missile strategic defence. The decision to decentralise the Soviet economic system and to some extent the military and political systems which has been noted in the past five years or so will no doubt continue. Kremlin military experts appreciate the major threats to Moscow, Leningrad, Kiev and other centres of control and for the next year or so they will not be able to meet the threat of devastating attack on these centres with any certainty. Thus a continuation of decentralisation, a strengthening of passive defence and a high priority to PVO anti-missile missile development will be major tenets of Soviet military and economic policy in the next few years.

The future composition and organisation of Soviet tactical air forces provide a major headache even for the most confident prophet. One can reasonably suppose that a much larger number of helicopters and air transport planes will be available to meet the extra requirements of mobility either in nuclear warfare or to meet the eventuality of operating ground forces in any part of the world at short notice. But one wonders to what extent the helicopter regiments will be integrated into the Soviet Army and to what extent the Soviet Air Force will develop vertical take-off supersonic fighters and fighter bombers. A major reduction in the size of the present force of single and twin-engined conventional tactical bombers is very likely in the 1960s as the Soviet armed forces increase the number of nuclear shells, rockets and bombs. New night and bad-weather aids such as infra-red and television will probably be in general use to support night and all-weather day attacks at any time of the year. The conventional campaign seasons of spring, summer and autumn have never rated as high in Soviet planning as with Western forces; with the new scientific aids they are likely to be less important still in the future. One wonders too if a force of helicopters will not be used for maintaining wireless and radio-telephone communications in duplicate and even triplicate to meet the immediate threat to normal communications which atomic weapons bring.

But one persists in thinking that the Soviet Defence Ministry will retain a large force of conventional fighters and bombers to meet the supply needs of her potential air allies and the future operational requirements of theatres of potential local wars in Africa, the Middle East and Asia, and even in the theatres of Europe where it may be considered that atomic guided weapons are of primary importance for quick nuclear neutralisation. But in the confused ground situation which is likely to develop the need to reconnoitre highly mobile and uncertain targets and to attack them individually and immediately may take precedence over any mass attack by batteries of nuclear artillery or rockets. That is why one imagines that the Kremlin will keep in being a force of several thousand conventional aircraft in tactical air units for at least the next five years or so.

But the big question which occupies both Moscow and Washington at the moment is the relative balance of the strategic deterrents in the East and the West and the likelihood of either side attempting a surprise or pre-emptive attack, because neither side has any reasonable prospect of success of effective defence against all-out nuclear attack. A bewildering series of appreciations have been published to suggest that in a year or so, the Soviet advantage in long-range missile attack will be so great that she will be tempted to carry out a surprise attack or a pre-emptive forestalling blow if for example there were to be a clash between the armed forces of Western and Eastern Germany.

The growing strength of Soviet long-range air and rocket attacks has certainly been fully appreciated and perhaps over-appreciated in U.S. defence circles in the past two or three years. In 1958 President Eisenhower warned in a broadcast: "Unless we act wisely and promptly we could lose the capacity to deter attack or defend ourselves." Early in 1960 General Power, Commander-in-Chief of the U.S. Strategic Air Force, observed pessimistically that "in 1962 we will have no active defence whatsoever against Russian missiles". He went on to estimate that in the current state of development of Soviet rockets, three intercontinental ballistic missiles fired at an unhardened air or rocket base would have a ninety-five per cent chance of destroying its target and that with 300 such successful rockets, of which half would be intermediate-ranged and half intercontinental in range, the Russians "could almost wipe out our entire nuclear strike capability within thirty minutes". Nor was this an isolated appreciation. At the end of 1958, the Rand Corporation, a United States government subsidised group of scientists working for the Pentagon defence authorities on strategic problems came to a similar conclusion. They estimated that in 1960 the USSR would have 300 intercontinental ballistic missiles of which at least 150 would be able to land within three miles of the target and the Rand Report warned that the Soviet long-range missile capability opened the possibility of being able to eliminate a high proportion of the Western deterrent, including the U.S. Strategic Air Command, in a single blow.

Let us assume the worst case during the next year or so. This is the case in which the USSR would have doubled or even trebled the force of 300 intercontinental ballastic missiles supported by more than 2,000 or more intermediate range ballistic missiles, and that in addition to these weapons the USSR has many thousands of tactical or medium or long range bombers which can add to the power of the Russian atomic potential and are intended mainly for a surprise assault on U.S. overseas bases, aircraft carriers and submarines. Let us assume a high degree of accuracy in their-long range rockets, a detailed plan for the immediate evacuation of the main Soviet cities, for the further dispersal of industry and the armed forces and maximum preparations for civil passive defence. Can they hope to organise on unalerted U.S. and NATO bases a mass attack which is bound to succeed? It is hard to believe that Soviet scientists, military leaders and politicians have not issued a series of caveats, some of them of considerable dimensions.

Not the least would be the possibility of an intelligence leak. The Soviet plan must be to launch a mass attack simultaneously. Such a plan will have to be constantly modified in great detail as new batteries of long range rockets become available and less reliance is placed on manned bombers and more on the rôle of Soviet submarines. No one would deny the high security standards reached in the Soviet armed forces. But the events of World War II must have taught the USSR some bitter lessons about the insecurity of even their high standards of security. The story of Igor Gouzenko the cipher clerk in the military intelligence department of the Soviet Embassy in Ottawa in 1945 is but one illustration of the obvious fact that the Kremlin cannot rely on the West not getting intelligence of a surprise attack. It must also be clear to them that they would have to communicate from day to day and week to week with their rocket and bomber forces if only to bring them up to date on the latest information about the Western deterrents which would involve changes in the target systems of Soviet rocket and bomber forces. With the experience and knowledge they have of the ease with which one or two clues in communications can either reveal a plan or at least alert an enemy, they cannot reasonably assume that an intelligence leak is not possible or probable, and they can have no means of knowing the degree of probability. One or two careless coders, a dissident official in high places, a perceptive but unsympathetic Pole, Hungarian or Roumanian in contact with Soviet rocket forces could be sufficient to blow their security scheme sky high. The Soviet authorities must also surely realise that in encouraging more visitors, including scientists and travellers of all kinds, to come to the USSR they are increasing the likelihood of such leakages of intelligence. Increased flights by foreign passenger planes and passengers all over the Soviet Union could also reduce the potential security of any Soviet military plan. Any cessation of these facilities after some years of détente in this field would be a long-term alert for the West.

A second and important factor which must cast doubt on any surprise attack plan is the difficulty of calculating in advance the probability of the success of any first strike against U.S. air and rocket bases. The view of many scientists and mathematicians is that the results of attacks with even thoroughly tested long range missiles cannot be accurately predicted. There is a high probability of great variation in the field performance of Soviet long-range rockets which can defeat any simple probability formula intended to guarantee that the USSR can with certainty eliminate, say, half the U.S.A. and NATO deterrent potential. It is no doubt simple enough to make a staff study calculation that so many hundreds or thousands of missiles and bombers launched simultaneously would give the USSR say a probability of nine chances in ten that seventy to eighty per cent of the Western deterrent would be eliminated in the first strike. But there are so many human and technical factors to disrupt the calculations of the military engineers and physicists. Can the Kremlin really believe that when the button is pressed all their aircraft and missiles will be ready to strike? Can they resort to a plan which could be thwarted if one submarine commander or one isolated battery commander gives an order to fire prematurely? And in any event surely the day and the time must be fixed many months in advance and be confirmed by some last-minute signal to all concerned. Surely even the high amount of extra communications traffic to implement such a plan would in itself be a warning to the West. An even more serious aspect is the problem of preparing liquid-fuel rockets to be fired in the mass at a given moment. To ensure maximum simultaneous firing from all points, they must be fuelled well in advance and this is likely to decrease the reliability of the rockets to a degree which is difficult for Soviet scientists and engineers to calculate.

A third factor which must give the Kremlin reason to pause is the difference between the demonstration of so called effective weapons and their performance in the field. The past demonstrations of the use of Soviet airborne troops, of anti-tank aircraft and other weapons at military manœuvres must have left Soviet leaders with no doubt in their mind that under the pressure of war conditions the theoretical performance of weapons deteriorates rapidly. No doubt there have been impressive demonstrations of Soviet long-range weapons in which élite engineers and military personnel have been employed and have demonstrated the reliability and accuracy of Soviet long-range weapons. But the question of the deterioration of performance and degradation of effectiveness when missiles are operated in the field must surely be uppermost in the minds of Kremlin planners. Even a reckless Soviet maker of decisions would know that the extent of the deterioration cannot be planned in advance. No doubt larger and larger margins of allowance can be made as Soviet rockets increase in range, reliability and numbers but the basic doubts must remain and the caveats will increase as the U.S. hardens her rocket sites, secures better dispersal for bombers and increases the number of her submarine-launched rockets.

Finally the USSR cannot reasonably evaluate the importance of the factor of the Western early warning system. They know that land-based radar has the potential of giving some 15-20 minutes' warning of impending rocket attack in a situation in which a Soviet mass surprise attack is set in motion. What in practical terms does this mean? How many U.S. and British bombers for instance can be airborne within 15-20 minutes as a result of this long-range radar warning which would show up immediately on radar screens in mass simultaneous attack conditions? How many could be effectively intercepted before they reach the target area? What is the effect on Soviet air defence of Western low-level attack, the use of the stand-off bomb and of radio counter-measures against Soviet radar?

To sum up, despite all the security precautions the Soviet may take and all the margins of error that their missile and bomber capability is extended to cover above the likely extent of operational requirements, there must always be large areas of doubt in the Kremlin about the success of any surprise or pre-emptive attack policy. Unless their intelligence is very bad, they must know they can stand up to a surprise attack from the West much better than the West can face a Soviet surprise attack and that such an attack from the West is, to put it mildly, most unlikely. The Soviet must also be as aware as the West of the danger of the uncontrolled and irresponsible behaviour of other powers, such as Communist China, the United Arab Republic and Eastern or Western Germany if an uncontrolled arms race continues with more and more countries being equipped with atomic weapons. And later, as weapons become more powerful and more diverse, the chances of psychotic and irrational behaviour by other powers will increase.

In a speech delivered in the House of Commons on March 1st, 1955, Sir Winston Churchill said: "Unless a trustworthy and universal agreement upon disarmament, conventional and nuclear alike, can be reached and an effective system of inspection is established and is actually working, there is only one sane policy for the free world in the next few years. That is what we call defence through deterrents. This we have already adopted and proclaimed. These deterrents may at any time become the parents of disarmament, provided that they deter." I believe that this statement made in 1955 is even truer in 1960 and that unless the U.S. fails to take the necessary measures to raise her deterrent forces, the weapons of mass destruction will turn out to be parents of an eventual international effective system of disarmament. This is not the moment to lower one's guard or to forecast anything but the fiercest economic, cultural and political competition between the Western and Communist worlds in the second half of the twentieth century. It is on the other hand almost certainly the moment to take Soviet disarmament proposals seriously and to have a step-by-step plan for seeing that in each phase of the disarmament procedure there are safeguards for all concerned. Whatever large-scale disarmament proposals are agreed to, the implementation of the proposals can only be achieved at a cautious and therefore slow rate. The ever-quickening pace of military weapon development, the increase of nationalism in Africa, Asia and Latin America, the unsolved problem of Germany and the potential rapid diffusion of the new military weapons to many countries give neither the East nor the West any alternative plan to disarmament which has a rational political or military basis. The only way for the Kremlin to ensure the long-term military behaviour of China and Japan is by effective disarmament machinery initiated in the 1960s. The same applies to the long-term behaviour of Germany in which the West is so heavily involved. As Marshal of the R.A.F. Sir John Slessor remarked more than five years ago, the weapons of mass destruction have made the abolition of large-scale wars almost inevitable. My belief is that Soviet defence experts will be asked to write papers on disarmament in the next few years on a greater scale than in the past and that if the disarmament talks are not bedevilled by mutual suspicion and an excess of propaganda, the second half of the twentieth century could become the first real epoch both of major discoveries and of peace. The alternative situation is two great power blocs with hands uncertainly and unhappily on the trigger, eventually losing the power to police their own allies.

BIBLIOGRAPHY

The Soviet Navy, ed. M. G. Saunders. Praeger (New York) and Weidenfeld and Nicolson (London).

The Soviet Army, ed. B. Liddell Hart. Praeger (New York) and Weidenfeld and Nicolson (London).

Soviet Air and Rocket Forces, ed. Asher Lee. Praeger (New York) and Weidenfeld and Nicolson (London).

Russian Aircraft. Ian Allan Ltd. (London).

Science and Technology in Contemporary War, G. Pokrovsky. Atlantic Books (U.S.A.) and Stevens & Sons (London).

Air Power, Asher Lee. Praeger (New York) and Duckworth (London).

War and the Soviet Union, H. S. Dinerstein. Atlantic Books (U.S.A.) and Stevens & Sons (London).

Soviet Strategy in the Nuclear Age, Raymond L. Garthoff. Atlantic Books (U.S.A.) and Stevens & Sons (London),

Other sources included the annual *Janes All the World Aircraft* and the monthly *Royal Air Force Flying Review*, both excellent sources and published in London.

INDEX

Index

Index